WORKING E

Page 89
instances of
spaces of 'home'

one of the chief features
of Linda's writing
dramatic clarity of
explanation
opens up windows
allows one to take
in fresh directions
oneself . . .

of
childcare

Studies in Urban and Social Change

Published

Working Bodies: Interactive Service Employment and Workplace Identities
Linda McDowell

Networked Disease: Emerging Infections in the Global City
S. Harris Ali and Roger Keil (eds)

Eurostars and Eurocities: Free Movement and Mobility in an Integrating Europe
Adrian Favell

Urban China in Transition
John R. Logan (ed.)

Getting Into Local Power: The Politics of Ethnic Minorities in British and French Cities
Romain Garbaye

Cities of Europe
Yuri Kazepov (ed.)

Cities, War, and Terrorism
Stephen Graham (ed.)

Cities and Visitors: Regulating Tourists, Markets, and City Space
Lily M. Hoffman, Susan S. Fainstein, and Dennis R. Judd (eds)

Understanding the City: Contemporary and Future Perspectives
John Eade and Christopher Mele (eds)

The New Chinese City: Globalization and Market Reform
John R. Logan (ed.)

Cinema and the City: Film and Urban Societies in a Global Context
Mark Shiel and Tony Fitzmaurice (eds)

The Social Control of Cities? A Comparative Perspective
Sophie Body-Gendrot

Globalizing Cities: A New Spatial Order?
Peter Marcuse and Ronald van Kempen (eds)

Contemporary Urban Japan: A Sociology of Consumption
John Clammer

Capital Culture: Gender at Work in the City
Linda McDowell

Cities After Socialism: Urban and Regional Change and Conflict in Post-Socialist Societies

*Out of print

Gregory Andrusz, Michael Harloe, and Ivan Szelenyi (eds)

The People's Home? Social Rented Housing in Europe and America
Michael Harloe

Post-Fordism
Ash Amin (ed.)

The Resources of Poverty: Women and Survival in a Mexican City*
Mercedes Gonzalez de la Rocha

Free Markets and Food Riots
John Walton and David Seddon

Fragmented Societies*
Enzo Mingione

Urban Poverty and the Underclass: A Reader*
Enzo Mingione

Forthcoming

Place, Exclusion, and Mortgage Markets
Manuel B. Aalbers

Cities and Regions in a Global Era
Alan Harding (ed.)

Urban Social Movements and the State
Margit Mayer

Locating Neoliberalism in East Asia: Neoliberalizing Spaces in Developmental States
Bae-Gyoon Park, Richard Child Hill, and Asato Saito (eds)

Worlding Cities: Asian Experiments and the Art of Being Global
Ananya Roy and Aihwa Ong (eds)

Fighting Gentrification
Tom Slater

Confronting Suburbanization: Urban Decentralization in Post-Socialist Central and Eastern Europe
Kiril Stanilov and Ludek Sykora (eds)

Social Capital Formation in Immigrant Neighborhoods
Min Zhou

WORKING
BODIES

INTERACTIVE SERVICE EMPLOYMENT AND WORKPLACE IDENTITIES

Linda McDowell

WILEY-BLACKWELL

A John Wiley & Sons, Ltd., Publication

This Edition first published 2009
© 2009 Linda McDowell

Blackwell Publishing was acquired by John Wiley & Sons in February 2007. Blackwell's publishing program has been merged with Wiley's global Scientific, Technical, and Medical business to form Wiley-Blackwell.

Registered Office
John Wiley & Sons Ltd, The Atrium, Southern Gate, Chichester, West Sussex, PO19 8SQ, United Kingdom

Editorial Offices
350 Main Street, Malden, MA 02148-5020, USA
9600 Garsington Road, Oxford, OX4 2DQ, UK
The Atrium, Southern Gate, Chichester, West Sussex, PO19 8SQ, UK

For details of our global editorial offices, for customer services, and for information about how to apply for permission to reuse the copyright material in this book please see our website at www.wiley.com/wiley-blackwell.

Library of Congress Cataloging-in-Publication Data

McDowell, Linda, 1949–
  Working bodies: interactive service employment and workplace identities / Linda McDowell.
    p.   cm. – (Studies in urban and social change)
    Includes bibliographical references and index.
    ISBN 978-1-4051-5977-7 (hardcover : alk. paper) – ISBN 978-1-4051-5978-4 (pbk. : alk. paper)
  1. Service industries. 2. Sexual division of labor. 3. Body, Human. I. Title.

HD9980.5.M3885 2009
331.7′93–dc22
                                                                    2009006549

A catalogue record for this book is available from the British Library.

Set in 10.5/12pt Baskerville by SPi Publisher Services, Pondicherry, India
Printed and bound in Malaysia by Vivar Printing Sdn Bhd

1   2009

Contents

List of Illustrations

Figure

Tables

Plates

Series Editors' Preface

The Blackwell *Studies in Urban and Social Change* series is published in association with the *International Journal of Urban and Regional Research*. It aims to advance theoretical debates and empirical analyses stimulated by changes in the fortunes of cities and regions across the world. Among topics taken up in past volumes and welcomed for future submissions are:

- Connections between economic restructuring and urban change
- Urban divisions, difference, and diversity
- Convergence and divergence among regions of east and west, north, and south
- Urban and environmental movements
- International migration and capital flows
- Trends in urban political economy
- Patterns of urban-based consumption

The series is explicitly interdisciplinary; the editors judge books by their contribution to intellectual solutions rather than according to disciplinary origin. Proposals may be submitted to members of the series Editorial Committee:

Neil Brenner
Matthew Gandy
Patrick Le Galès
Margit Mayer
Chris Pickvance
Jenny Robinson

Preface and Acknowledgements

This book is both a review of some of the exciting case studies and ethnographies of waged work in the service sector that have appeared in the last two decades or so and a synthesis of some of my own work in the area. Its focus is on jobs that demand the co-presence of the provider and the purchaser of a service, jobs in which the embodied identities of workers are on display and affect the exchange in ways that are uncommon in manufacturing employment. Embodied performances are an essential element of everyday interactions in service sector workplaces. It is a book about the associations of the body with deference, shame, disgust and dirt, as well as with beauty, health and fitness, adornment and decoration and the ways in which these characteristics and attributes become associated with differently gendered, raced and classed bodies, and so are rewarded or penalized in different ways within the labour markets of advanced industrial nations. The main, although not sole, focus is workplaces in Great Britain, although there are also examples drawn from US labour markets and workplaces. Furthermore, global migration means that even the most local of workplaces in a particular nation-state may include employees from a wide range of countries. It is a book about socioeconomic change, as well as migration, about the growth of low-status and often precarious forms of work and their distribution across what geographers term spatial divisions of labour: the places where work takes place and the connections across space that produce their specificities.

I want the book to be useful as a source for students and their teachers: it provides a set of case studies that can be dipped into as well as, or instead of, being read in its entirety. The first three chapters, however, are essential reading as they lay out the empirical and theoretical bones of the argument about the significance of interactive service employment. In the chapters that follow I discuss a range of case studies or ethnographies of different types of what I define as high-touch interactive work. In these chapters I explore not only the findings of the different studies but also discuss how they were carried out. This focus on methods should help those contemplating their own research on particular types of work as well as provide some ammunition to answer those critics of qualitative methods who argue that case-study approaches are 'too anecdotal' or 'not representative'. The bibliography is large, providing details of a large number of studies which should lead to yet more.

Many of my discussions about service sector employment have taken place in lectures and classes for a course on the geographies of work and employment that I have taught over the last decade in Cambridge, London and Oxford. I am grateful to all the students who have taken the course or done their doctoral work with me, sharing their experiences of the labour market as both researchers and temporary employees. Their working lives and the range of jobs held in vacations were often much more varied than mine and helped me think harder about what happens in daily life at work. Four Oxford students – Ian Ashpole, James Cohen, Kat Hardy-King and Laura Putt – read the book just before their finals in summer 2008 and Sarah McDowell did the final proofreading. Special thanks to them. I should also like to thank old and new colleagues for making each of the departments that I have worked in both enjoyable and distinctive environments, each with its own particular working culture.

Some of the chapters are partly based on revised versions of articles that I have previously published and draw on work supported variously by the Economic and Social Research Council, the Joseph Rowntree Foundation and the British Academy. I thank these funding bodies for their support. The work in a hospital and a hotel that is part of the subject matter of chapters 7 and 8 was undertaken with Adina Batnitzky and Sarah Dyer and I am extremely grateful to them not only for their input into the research but their willingness for me to include aspects of it here. Particular thanks are due to the Leverhulme Trust who granted me a two-year Major Research Fellowship between October 2006 and 2008. That award gave me the space and time to read and think and then to write this book. Two anonymous reviewers read the manuscript at a crucial stage and their tough but invaluable comments have made this a better book. If I knew who you were, I would thank you in person. Finally, Neil Brenner, editor of the series and Jacqueline Scott, at Wiley-Blackwell, were always supportive at just the right times.

The author and publisher would like to acknowledge the editors and publishers for permission to reproduce here in revised versions sections from papers in the following journals: *Antipode* 2006, 38 (4), pp. 825–50; *Economic Geography* 2007, 82 (1), pp. 1–28 and *Progress in Human Geography* 2008, 28 (2), pp. 145–63. While every effort has been made to contact copyright holders for permission, the publishers would be grateful to hear from anyone who is not acknowledged and will correct this if a second edition of the book appears.

Linda McDowell
Oxford

1

Service Employment
and the Commoditization
of the Body

It is illiberal and servile to get the living with hand and sweat of the body.

William Alley, *The Poore Man's Library*, 1571

Work is no less valuable for the opportunity it and the human relations connected with it provide for a very considerable discharge of libidinal component impulses, narcissistic, aggressive and even erotic, than because it is indispensable for subsistence and justifies existence in a society.

Sigmund Freud, *Civilization and Its Discontents*, 1930

Introduction: Narratives of Change

This is a book about bodies at work. It is about who does what sort of waged work in the contemporary service economies that dominate the western world. Its aim is to explore who does what, where, with whom, for whom and to whom, and with what consequences for the financial rewards and social status that accrue to different workers. Its focus is on the social division of labour, on the ways in which class, gender and ethnicity, as well as age, looks and weight, are key attributes in explaining who is employed in what sorts of work in the first decades of the new millennium. The types of work discussed are those where both the worker and the consumer are present and, in the main, where the service provided is used up at the time of the exchange. Through the lens of the workplace, the emphasis is on the sorts of embodied interactions that take place in everyday exchanges between the three sets of actors involved in these exchanges: workers, managers and clients/customers. As a consequence, the book is about a smaller and more local spatial scale than geographers are used to. Traditionally, economic geographers' scale of analysis is local or regional labour markets, although there is also a strong geography of the firm. Sociologists are more typically analysts of workplace interactions, yet they too often ignore the significance of the local, neglecting to ask what is specific and

different about the places they study. The aim here is to bring a geographical
and sociological perspective together. Although the focus is the workplace – be
it an individual home, the street, a shop or a hospital – through this lens, the
changing national and international spatial divisions of labour that produce
increasingly diverse workforces in the cities of western economies are also
revealed. The examples are drawn in the main from UK workplaces, as well
as from the USA, but the workers whose identities are at the heart of the book
are extremely varied in their national origins.

In the new millennium in the economies of the western world, it may seem
as if we have entered a new era or at least have been passing through a
period of significant change since 'around 1973': the date that David Harvey
(1989) identified as the start of the transformation from the old Fordist model
of economic organization to a new condition of postmodernity. Since then,
growing numbers of individuals have become part of the social relations of
waged work, and many seem to be working in new types of work under different conditions compared to workers in a former era, especially those in
manufacturing industries. Change is the key motif of contemporary discussions of waged work. In both the popular and academic literatures about
waged work and the labour market titles that indicate a radical change with
the past are common. The 'new capitalism' (Sennett 2006), the rise of 'post-
Fordism' (Amin 1994), in a period of 'liquid modernity' (Bauman 2000) are
all part of the titles of books purporting to describe the changes that are
evident in what some have identified as a 'new' economy (Carnoy 2000;
Jensen and Westenholz 2004). 'One of the most striking things about much
of contemporary theorizing about work and identity is the epochalist terms
in which it is framed', in which the 'logic of dichotomization establishes the
available terms in advance' (Du Gay 2004: 147).

The statistical picture of labour market change seems to confirm these claims
about transformation: from the Fordist manufacturing economy of the past, to
a post-Fordist, post-modern or post-industrial service economy of the present.
There has been a remarkable shift in the types of work most people in the UK
perform (and in the USA, France, Germany and other western economies),
as well as in the types of people who fill the expanding jobs. The most significant change has been the growing dominance of service employment: almost
three quarters of all employees in the UK currently work in services. Half-way
through the twentieth century, more than half of all workers were employed in
the manufacture of goods; half a century later, the sons and daughters of these
workers do quite different types of work. Less than 15 per cent of British workers are now employed in manufacturing. This shift has been captured in terms
such as 'deindustrialization' (Gregory and Urry 1986), the 'end of manufacturing', even, for some commentators, 'farewell to the working class' (Gorz 1982),
as it seemed to these analysts almost inconceivable that working in shops or
cafés, providing massages, cleaning houses and teaching children was 'work' in

the old sense of producing material products through the application of brute strength and heroic effort, typically by men. Other commentators have seen the rise of service employment in more optimistic terms, welcoming the introduction of less stressful or dangerous working conditions, improved pay and opportunities for social mobility for a better-educated population (Wright Mills 1953; Bell 1973; Gartner and Reissman 1974). By the new millennium the optimistic scenario had shifted into a different register as it became clear that the service sector includes poorly paid, as well as well paid, jobs and that for many the anticipated social mobility would never materialize. To narratives of loss and nostalgia for the golden age of manufacturing – a period from about the end of the Second World War until the 1970s – a new narrative of precarious and insecure work was added (Sennett 1998, 2006).

Associated with this apparent transformation of the labour market and adding to the claim of many that older forms of working-class solidarity are dead, there has been a marked feminization of the labour force in Western Europe and the USA. Rising numbers of women have entered the social relations of waged work over the last fifty years, many of them in the service sector and many on different terms and conditions than those that typified masculine forms of attachment to the labour market. Many women are employed on a part-time basis, especially in the UK, and tend to work in different sorts of workplaces than the factories, shipyards and car plants that typified male work in the Fordist era. Women are more likely, for example, to work in other people's homes or in small, often unorganized, workplaces. And as well as growing numbers of women, economic migrants – people born outside the nation-state where they labour – are filling a growing number of jobs in service economies. Thus, labour forces are becoming increasingly diverse in their social characteristics.

At the same time as waged work is changing, and often becoming more precarious, it has become a more central part of more and more people's everyday lives in almost all countries in the global economy, apart from the former communist countries where participation rates fell in the 1990s. Participation rates in the UK labour market, for example, are higher than at any previous period. In 2008, 75 per cent of all people of working age in the UK were employed: one of the highest participation rates in Europe. Two decades earlier, British sociologist Ray Pahl (1988) argued that employment – the nature, distribution and rewards for waged work – was one of the most significant and urgent issues for social scientists to address. He suggested then that a range of questions about work remained unresolved as 'confusion and ambiguities about its meaning, nature and purpose in our lives are widespread' (p. 1). His argument is even more relevant now, despite a huge expansion of academic analyses of waged work. As the economist Francis Green (2006: 1) argued, 'almost everyone gets to do it. Work itself is a major and defining part of most people's lives. It takes up a large proportion

of their time on this earth and profoundly moulds their life experiences.'
Developing an understanding of how labour markets work, how people are
divided between and segmented into particular jobs, and what this means for
their self-identity and standard of living, is an even more significant task as
the new millennium advances and the place of waged work in most people's
lives continues to expand.

Over the last two decades the expansion of employment has also become
a central part of the reforming agenda of central and local governments.
The British state, under New Labour governments since 1997, has endeav-
oured to increase the proportion of the national population that is economically active. In common with other western governments, albeit to different
degrees, the adoption of a series of neoliberal employment policies has
resulted in cuts in state provision and benefits for the workless and a correspondingly greater emphasis on the responsibilities of individuals to provide for themselves rather than rely on state support (Herod 2000; McDowell
2004; Peck 2000). Growing numbers of citizens in these countries – and in
countries where systems of state support for the workless have never been
well developed – are now expected to provide for themselves though labour
market participation and are doing so by entering into the social relations of
waged work in unprecedented numbers. The UK and the US are among
the most 'liberal' economies, where the dependence on deregulated markets
and workfare policies distinguish them from mainland Western European
economies and from the Scandinavian countries (Esping Andersen 1990, 1999;
Perrons et al. 2006), which provide greater protection for the majority of
workers. In the UK and the USA 'workfare' programmes have replaced
parts of older forms of welfare provision (Peck 2000; Sunley, Martin and
Nativel 2001) which time-limit access to benefits as well as enrol the workless
onto 'job ready' programmes. In the UK there is also a move to reduce the
eligibility for benefits of long-term sick and incapacitated claimants: an issue
that is particularly important in parts of the country where men previously
worked in heavy primary or manufacturing occupations (Fothergill and
Wilson 2007). Young unskilled men and women, not only in these regions
but across the UK economy, are also 'encouraged' into work through
schemes designed for their age group. Single parents – the majority of whom
are women – are a further target group for raising participation rates which
currently are well below those of other women workers.

For the more affluent and highly skilled groups in the population, apart
from men over 50 whose participation rates are falling, employment participation rates have continued to rise in the last two decades. As they do,
time becomes more precious for waged workers, exacerbating the transfer
into the market of many of the services previously provided within the
home. This is one of the most significant changes in the way in which consumer services are provided in advanced capitalist economies. Rather than

the rise of a self-service economy, identified by Gershuny (1978), in which households serviced their own needs in the home through the purchase of consumer durables, more and more services have become part of the market. As well as rising attendances at cinemas and in other leisure spaces, dual-income middle-class families, for example, increasingly buy childcare, house cleaning, dog walking and a whole gamut of other services that enable their complex lives to function more smoothly. The demand for these low-waged jobs in turn draws less affluent individuals – typically, women – into the labour market (Sassen 2001), often re-cutting class and gender relations, as I shall illustrate in the case-study chapters. Working-class employees in these low-wage 'servicing' jobs increasingly work in intimate contact with the bodies of the middle class in ways that are more reminiscent of earlier periods of industrial history than the immediate Fordist past (Panitch and Leys 2000).

The initial response by labour economists, geographers and sociologists to manufacturing decline in the 1970s and 1980s was to focus on the causes and consequences of deindustrialization, rather than analyses of the causes and consequences of this growing commoditization of services. A range of studies addressing, for example, questions about declining wage rates and the loss of the middle of the income distribution (Bluestone and Harrison 1982), diminishing union membership (Martin, Sunley and Wills 1994, 1996), male job loss and regional inequalities (Gregory and Urry 1986; Massey 1984; Massey and Meegan 1982) dominated the agenda of labour market studies. Some argued that the decline of manufacturing heralded the 'end of work' (Rifkin 1995), whereas others suggested that the rise of a consumer economy was leading to a reduction of the significance of employment in the social construction of identity, as people increasingly constructed their sense of self-worth primarily through consumption rather than through work (Bauman 1998): perhaps surprising claims as governments pushed through workfare programmes.

More recent studies of economic change and its labour market correlates, however, have focused explicitly on service sector employment growth and its meaning for workplace identities as well as for national economic growth. The dominant emphasis in this literature has tended to be an optimistic one: a 'new' knowledge economy (Carnoy 2000; Rodrigues 2005) was identified based on the production of 'weightless' goods and services such as financial commodities, legal expertise and new forms of scientific innovations in an economy that is 'living on air' (Coyle 1997; Leadbetter 1999). In this weightless economy, an increasingly specialist and well-educated workforce, employed in new clean environments, facilitates successful competition with newly industrializing countries in a globalized world, permanently connected by the internet (Castells 2001). National economies are linked into a global 'space of flows' connected by new technologies that facilitate the flows of information and capital (Castells 1996). But there is also another narrative of change that has not caught the attention of theorists or the media to the same extent. This story emphasizes the less successful and less glamorous side of service sector growth, recognizing the steady expansion of poorly paid, low-wage jobs, increasingly undertaken by women or by economic migrants but also by less-well educated or low-skilled men, especially young men. These are the workers who stepped into the gap to provide daily services to keep the households of work-rich but time-poor households running smoothly, as well as to care for the elderly and service the growing leisure demands of the majority. Here too a story of regret is evident, as commentators (Sennett 1998) mourned the loss of permanent, well-paid work for men, replaced by new forms of more precarious attachment (Vosko 2001) to the labour market in which employers have little loyalty to their workforce, which is both disposable and replaceable.

What unites these three narratives – of deindustrialization, the 'new' knowledge economy and the rise of precarious work – is the rhetoric of radical change or transformation. Western economies seemed to be being significantly restructured as new ways of working in new forms of employment replace the older certainties of full-time waged work, often for a single employer over a lifetime. I argue in the next chapter of this book, however, that the rise of service employment is neither new nor a significant transformation, or rather not a transformation that can be explained solely by studying the economy in isolation. Instead, I want to argue that what we are witnessing is the commodification of many of the types of work that were previously undertaken mainly in private homes and for 'love' – in the sense of not for wages. Because participants in these activities – the care of children, sexual relations, caring for elderly bodies, meeting leisure needs – were not financially recompensed, they were not recognized as work and so excluded from economic analyses. If they had been included, then not only would the definition of work have been widened but the attributes of both work and workers would have been different. Women performing caring labour in the home would have been included in this wider definition of work and so the apparently radical change captured in the dominant narratives would have seemed less remarkable. Here, instead of a narrative of transformation, I want to emphasize continuity. In part, I do this through the use of quotations at each chapter head from earlier eras in British labour market history which remain entirely relevant today. In this way the connections between the types of work and employment that dominated what are now regarded as past eras are connected to the continuing changes in the post-millennium labour market. Servicing the bodily needs of others has always taken up the time of large numbers of workers and typically it has been regarded as low-status work, and constructed as particularly suitable for women.

Servicing Work as Waged Employment

While I shall argue that continuity is the key defining characteristic of all
those forms of service sector work that involve caring for the bodily needs
of others, there undoubtedly has been a significant shift in the nature and
location of these types of work. In the long decades of the postwar years of
economic growth and relative prosperity, notwithstanding manufacturing
decline, these forms of embodied work have moved from the private into the
public arena (provided both by the market and the state) and into the cash
nexus. The rise in women's labour market participation rates and their
increasing inability, or unwillingness, to provide the range of services in the
home that other family members once took for granted has exacerbated this
shift. In affluent societies, an increasing array of activities, including sex,
food, healthcare, advice, exercise, childcare and music making, that was
provided for 'love' (in the sense of financially unrewarded) within the family
or local community, has become part of market exchange.

The movement of these services into the cash economy and women's
greater involvement in the social relations of employment, has had a significant impact on women's sense of self-identity, as well as on the location
of their key social relationships. For all workers, the workplace is an arena
of social relations, a place in which peers, superiors, subordinates, managers
and bosses interact. These sets of relations are not merely the rational
exchange of labour power for wages, as economists argue, but also provide
an outlet for emotions and the development of social relationships: aspects
of the employment relation more typically analysed by sociologists and psychologists. People experience anger and frustration at work, as well as satisfaction in a job well done. People make friends at work, sometimes fall in
love, and provide and receive emotional support, as the quotation from
Freud at the head of this chapter makes clear. Work is about sociability.
But what has changed in the twenty-first century is the growing significance
of social relations in the workplace in women's lives. Two generations ago,
for most women, socializing took place in the home, among family, kin and
neighbours, and in the local community, perhaps through participation in
voluntary work or through membership of clubs and groups based on
mutual interests. As the traditional saying made clear, 'a woman's place is
in the home'. In 2008 the majority of women are in waged employment
and so are able to construct an alternative set of social networks. Indeed, as
the US sociologist Arlie Hochschild (1997) found in her study of a large
Midwest workplace, women employees not only valued their workplace
relationships but regarded them as more significant than their home lives
for friendship and emotional support. As the subtitle of her study – when
'work becomes home and home becomes work' – indicates, for these women

life had changed as emotional satisfaction was sought and found in the workplace rather than or as well as at home. This change affects others too, not only at home, but in communal and voluntary patterns of participation. As Robert Putnam (2000) noted, many Americans are now condemned to go 'bowling alone'.

As well as social and emotional relationships with other workers, increasingly, participation in waged work in Britain, the USA and in other service-dominated economies involves the establishment of social relationships, albeit typically transitory, with customers. The most common forms of waged work now undertaken in the UK involve a service exchange between employees and customers – what Robin Leidner (1993) termed 'interactive work'. In the service sector, the production of consumer services typically depends on the co-presence of workers and purchasers – waiters serve restaurant customers, the hairdresser is only able to cut and style hair when the client is there, teachers need an audience for their classes, doctors a body to practise on. Further, the service is used up at the moment of exchange: a meal is over when customers leave, a haircut completed, a lecture finished, an operation over.

The recognition that service employment involves social interactions in spaces where the service provider and service purchaser are *co-present* is connected to a second shift in the last half-century: that is the huge expansion in the range and patterns of consumption in modern economies as an increasingly wide range of goods and services are for sale. As well as the typical goods of an expanding consumer economy – cars, fridges, clothes, an ever-wider range of electronic gadgets – there has been a remarkable expansion in consumer services, from eating out, through travel, to all sorts of care for the body and the soul, from massage to counselling. The economy as a whole has become one in which the provision and sale of services is the key motor of development as well as one in which a great deal of that emotional solace as well as bodily care that was previously provided in the private arena is now for sale. In these consumer-driven, service-dominated economies, individuals' bodies – as the producer and objects of exchange – have become absolutely central in a way that differs from the embodied labour power driving manufacturing industries. The physical attributes of the body providing a service are part of the exchange that occurs at the point of sale. A well-groomed, preferably slim body, produced through exercise, adornment and self-improvement, whether temporary through the application of cosmetics or more permanently through radical interventions such as surgery, is seen as an essential requirement of many, if not most, forms of employment. While this is not a new phenomenon – all types of work depend on a range of embodied attributes and visceral emotions ranging from physical strength through distaste and disgust to empathy and affection – it is significant in the extent to which it draws in ever-greater numbers of workers, including men, into forms of employment that depend on visible, interactive and embodied exchanges in which the physical shape of bodies, their adornment and workers' emotions matter in workplace performances.

This sort of work is more personal, centred within a set of direct contacts and social relations between providers and purchasers, than work in the manufacturing sector, where workers – the producers of goods – rarely have any contact with the eventual purchasers of what they make. In the service sector, the personal embodied attributes of workers enter into the exchange process in a direct way. When we ask for a waiter's advice about what to choose to eat, or consult a stylist about what might suit us, we evaluate their advice not just on the basis of the technical information that they might give us but also on the basis of what we think about them: whether we find them sympathetic or trustworthy, whether they are personable, friendly, standoffish, even aggressive, whether we admire or resent their youthful good looks or their facial piercings and fashionable dress. And so transactions in the service sector depend not only on a cash exchange but also on a personal interaction in which the embodied attributes of both provider and clients enter into the relationship, however momentary or transitory it might be. The embodied attributes of workers are part of the service – their height, weight, looks, attitudes are part of the exchange, as well as part of the reason why some workers get hired and others do not for particular sorts of interactive work. The exchange is also an emotional one in which the tastes, predilections and attitudes of both parties to the exchange are part of what is going on in the workplace, which is more likely to be a shop or a café, a nursing home or a childminder's house than a factory, shipyard or machine shop. These two concepts – embodiment and emotions – are central to understanding service employment.

Many of these expanding forms of interactive employment, for both women and men, increasingly involve working *on* the body. They involve servicing the bodies of consumers, clients and patients – in healthcare, in gyms, massage parlours and hair salons. As I have already noted, they also demand that employees work on their own bodies in self-improvement programmes to produce an idealized version of a slender, toned, deodorized, youthful-looking (and white) body – the type of body that is most highly valued in the new consumer-based economies of western cities. Images of these idealized bodies are ubiquitous. Where once representations of toned, white, often semi-clothed women were the typical images used to sell a range of products, in the twenty-first century there is greater gender equality in advertising. Men's bodies are also now used to sell products. Images of muscled men have escaped the covers of men's health and sports magazines, for example, and joined the more waif-like figures of the young men used to sell fashion. The male body is now an object of admiration, to be looked at and admired not only by straight women and gay men but also by heterosexual men (Simpson 1994) and so the male body has become a mainstream consumer icon, using to sell a wide range of good and services. Thus, bodies sell objects and services as well as perform services. And the bodies of those selling services are also expected to conform to these idealized images of masculinity and femininity that feature in advertisements. Men, as well as women, have to pay greater attention to issues about style and bodily presentation, to looking appropriate for work, to dressing for success and to maintaining the body. The ambiguity of the book's title is deliberate – 'working bodies' refers both to workers themselves going about their daily tasks in the labour market and to the work that they do on their own and others' bodies as part of the labour process.

The significance of embodiment is not new. All work depends on embodiment: a fleshy person has to turn up every day, whether to a coal face or an office. But the new emphasis on interactive embodiment is also restructuring class and status distinctions, as well as gender divisions of labour. There has long been a set of class distinctions in Britain based on the mind-body dichotomy and a belief that manual labour, embodied physical labour demanding brute strength, was a less valued way of earning a living than the application of thought. The shift from 'brawn' to 'brain' jobs, for example, is celebrated in the contemporary vision of a knowledge-based economy, where the trivial daily tasks of servicing the population are ignored. This mind-body: non-manual-manual distinction also maps – or used to – in large part on to gender divisions, although both brawn and brain typically are associated with a dichotomized version of masculinity. The apparently entirely cerebral demands of white collar or scientific work have long been associated with masculine attributes such as the capacity for logical thought and rational behaviour and, of course, this type of work positions the labourer in the middle class, dividing this worker from lower-status workers who rely on their bodily strength. But as feminists have argued, the disembodied rational middle-class male worker with no emotional needs or dependents, who apparently epitomized the Protestant ethic that initially ensured the industrial revolution and its bureaucratic management was an Anglo-Saxon affair, is more akin to an angel than a living man. Christine di Stefano (1990) once memorably commented that this idealized masculine being should be brought down to earth and given a pair of pants. All forms of employment, even the apparently least physical in terms of the energy expended, depend on an embodied, gendered worker applying effort in exchange for income. The types of hard manual labour that were dominant for so many centuries have, however, all but disappeared and far more people in the UK are now employed as care workers than as miners. Yet, paradoxically, as I demonstrate throughout this book, the significance of embodiment – of working by hand and sweat of the body – has increased. What has changed is the association of bodily labour with masculinity.

At the bottom end of the labour market, 'feminine' attributes of care and empathy are now constructed as ideal job attributes in interactive embodied work and it is men who may find that their embodied attributes are a disadvantage.

Theorizing Embodied Work

As the significance of service work has increased, there have been noticeable changes in the ways in which work and employment are theorized. It is abundantly clear in service occupations that labour is unlike other commodities: men and women enter their place of work as a set of living beings, opinionated, awkward, obstinate or deferential, embedded in sets of social relationships of kin and friendship. These characteristics as well as their age, ethnicity and gender, affect the ways in which they go about earning their living. At the same time, social exchanges and interactions with customers and clients affect as well as are affected by the attributes and identities of workers, influencing the ways in which people and the types of jobs that they do are differentially ranked and rewarded in the labour market. As a consequence, new ideas about the social construction of identity at work and workplace performances have become a part of theoretical debates about the changing nature of work.

Challenging Pahl's insistence on employment and the division of labour as the key issue for sociological analysis at the end of the twentieth century, Bryan Turner (1996), also a sociologist, made a counterclaim. He suggested that the body or embodiment was *the* crucial issue for sociologists as the new millennium approached. He argued that the body had become the key issue in social theory as contemporary consumer societies emphasize 'pleasure, desire, difference and playfulness' (p. 2). Like Bauman, Turner suggested that consumption rather than production was not only the new motor of the economy but also the critical site for the social construction of self-identity. But until relatively recently, sociologists of the body and sexuality ignored the labour market as a site of embodied performance, concentrating instead on leisure spaces and other sites of consumption in contemporary societies. However, they noted that in these locations work *on* the body became a central part both of self-production and maintenance. Desirable body shapes were achievable not only through self-improvement (dieting and exercising in the main) but also through new forms of surgical intervention as bodies are trimmed, tucked, reshaped and remade. Indeed, such is the ability to alter and transform the materiality of the body that Giddens (1991), Shilling (1993) and others have argued that it has become a significant site of uncertainty in the postmodern world, raising anxieties about the connection between embodiment and identity, as bodies become both mutable and, increasingly, virtual. I return to the significance of virtual bodies in the concluding chapter.

In feminist theory, the body and bodily emotions have also become a more central focus of scholarship in the last fifteen years or so. Indeed, as Fonow and Cook (2005: 2012) argued:

> Today it is hard to imagine a time when we did not know that bodies and their location mattered or that rationalities were gendered.... From the beginning, feminists challenged the artificial separation of reason (mind) and emotion (body), and they have come to view emotion as both a legitimate source of knowledge and a product of culture that is as open to analysis as any other culturally inscribed phenomenon. The significance and legitimacy of emotions as a topic of inquiry, as a source of theoretical insight, and as a signal of rupture in social relations is now well established in feminist circles.

The huge growth of work on embodiment was not anticipated, in part because feminists tended to resent the fleshy body, fearful of making it a central part of their theorizing in case it seemed as if biology was destiny. Instead, the initial aim of feminist scholarship was to decentre the significance of the material body, decoupling sex from gender and arguing that gender was a social construct rather than a reflection of biology. As Fonow and Cook (2005: 2012) note, 'it did not occur to us to view the physical itself as a social construct'. Theorizing the body as a social construct has, however, become central to more recent work, in which the body is seen not only as an object of analysis but also as a social category of inquiry. Feminists now talk about disciplining the body, sexing the body, writing the body as bodies are theorized as sites of culturally inscribed and disputed meanings, experiences and feelings that are both the source of theoretical insights and subjects of analysis. The body – at work, as an object of work and the role of emotions as a central element of work – has become a key part of feminist analyses of labour market change in service economies. Thus, in feminist analyses of employment, claims about the significance of employment change and about the importance of the body have been brought together to provide new understandings of work and employment.

If sociologists of sexuality and the body forgot, or at best under-emphasized, the social relations of employment, then theorists of work and employment tended to forget the labouring body and its sexual desires and fantasies. Apart from work by feminist economists, sociologists and geographers about embodiment, bodies, especially female bodies, were ignored by mainstream labour market theorists or at best relegated to the borders of their respective disciplines on the grounds that it was 'about women'. However, for more than two decades, feminist theorists have addressed questions about the significance of emotions at work and latterly questions more directly concerned with bodily performances and their relationship to emotions and affective relations in the workplace. In her book *Bodies at Work* (2006), which provides an excellent review of some of the recent sociological work about embodiment, Carol Wolkowitz showed how by crossing conventional boundaries between the sociology of work and the sociology of the body, a richer understanding of the nature of contemporary labour market change is possible. Wolkowitz built on decades of feminist theorizing that has challenged conventional analysts who ignored embodied differences between workers. Arlie Hochschild's *The Managed Heart* (1983) is now a classic text for all studies of how emotions are part of a commercial exchange and since then there have been numerous other excellent studies of emotional labour and embodied work – the two are not always coterminous, as I shall show in the case studies that follow – in sites ranging from massage parlours to the operating theatre, from call centres to university seminar rooms.

One of the most significant early developments in feminist theory that drew attention to the ways in which the associations between emotions/ femininity and rationality/masculinity structured waged, as well as unwaged, labour in the home, was the concept of the gender or gendered division of labour. This concept allowed a forceful critique of the masculinist bias in both employment itself and in theoretical explanations of its nature and consequences that, as I indicated above, influenced the ways in which deindustrialization and the expansion of service employment were analysed. Feminist scholars pointed out that the gendered attributes of workers were a key part of understanding whether and how people became waged labourers and what rewards accrued to them once they did so. Through theoretical innovation and careful empirical work, the ways in which men and women are unequal in their access to and rewards from waged work were investigated. The gendered nature of the public and private spheres was documented, showing how a set of ideological beliefs about women's place in the home and domestic sphere resulted both in their exclusion from the labour market or/and their segregation into suitably female/feminine positions. Work fit for women was then regarded as less significant than either waged work undertaken by men or than women's unpaid labour within the family. Male breadwinners were expected to provide for their dependents through waged work and women through their unpaid labours in the home. If women did enter the labour market, they were regarded as less skilled and so paid a lower wage than men. As women workers self-evidently were seen to rely on their 'natural' talents (variously defined as dexterity, empathy, a caring heart) at work, then it was clear that they should not expect equivalent rewards to men, whose skills had been honed by experience or training. Thus women have laboured for most of the last hundred years or so in explicitly 'female sectors' of the labour market for lower pay than men (Bradley 1989). Even in 2008 in the UK, when women are more than half the university intake and are making inroads on many of the professions, a gender pay gap between men and women working on a full-time basis still exists: women earn just under 88 pence for every £1 that a man earns and the gap is far wider for women who work on a less than full-time basis.

Feminist scholars have also shown how the sets of social attributes that are mapped onto gendered bodies become crucial in constructing divisions of labour and hierarchical evaluations of worth. Embodied characteristics such as skin colour, weight or height, accent and stance map onto gender to produce a finely graded set of evaluations that position workers as more or less suitable to perform different types of work and different sets of tasks. The production of this hierarchy of suitability is a key part of the case-study chapters and I expand on its theoretical significance in chapter 3.

Place Matters

As well as gender divisions of labour, there is a visible spatial division of labour. The social characteristics of workers and the types of work they do vary across geographic space at different scales of analysis. There is now a small but growing literature about the geographies of labour (Herod 2001; Wills and Waterman 2001; Hale and Wills 2005) and the spaces and places in which work takes place (Peck 1996; Castree et al. 2004), as well as commentaries addressed to other disciplines about why space matters (Castree 2007; Herod et al. 2007; Ward 2007). In sociology, workplace-based studies have always been more common than in other social science disciplines and there is a long tradition of ethnographies and case studies of particular types of jobs and occupations. Most of these studies, however, neglect the specificity of place and tend to ignore questions about how workers are assembled, why they differ from workers in other places doing the same sorts of work. Indeed, sometimes (often for reasons of confidentiality) sociologists change the name of their workplaces or move them to different cities (Cavendish 1982; Glucksmann, 2009), and so the specificities of local variations in the supply and demand for labour and the difference it makes if a factory or hospital is located in say Birmingham rather than in London cannot be addressed. In the case-study chapters I include questions about the signifi- cance of place: are workers from the immediate locality? If not, why not and where are they from? What difference does place make to workplace practices? Although interactive embodied work is by definition the most local of all sorts of work – the provider and client must be co-present in the exchange – it is not necessarily provided by local workers.

Clearly, however, because of co-presence, the very local scale matters in service economies. All service exchanges take place somewhere – in the home where wives and mothers provide for love the same sorts of services that nannies, cooks and cleaners provide for remuneration in affluent or time-pressed households, and in numerous specialist locations, including shops, factories, banks, universities and schools, parks, playgrounds, gyms and gardens. Work for wages now also takes place in cars, on trains, in parks and cafés, as new forms of technology enable workers to free themselves from the fixed boundaries of their workplace. But paradoxically these same technologies are mechanisms of social control as they enable employers to increase their surveillance of employees and intrude into the spaces once solely associated with leisure or with social relations based on affection and mutual obligation rather than the cash nexus. Nannies working in the 'private' spaces of the home, for example, may be subject to covert forms of electronic surveillance; call centre workers and data processors may have key strokes counted and in all sorts of workplaces CCTV cameras increasingly monitor workers' movements as well as deter or record potential intruders. The scale of the immediate workplace is the focus here, but even this most local of foci demands the analysis of processes at a larger spatial scale: the city as a whole, the region and the nation-state, as well as the analysis of the changing pattern of connections between these different scales, whether measured by flows of capital, goods, information or labour. As I show in later chapters, workers in many of the most intimate forms of embodied work are assembled across a broad spatial canvas, sometimes on a global scale, although in ways that particular systems of national and local regulation influence, so constructing new hierarchies of eligibility. The intersection of spatial processes at different scales produces in local labour markets the specificities of particular types of work and workplaces.

Location at this wider scale – within and between regions and within and between nation-states – clearly affects what types of work people do, as not only are there geographical differences in the distribution of particular types of employment but also in the numbers of jobs available in particular parts of the country, in different sized settlements or different countries. Individuals searching for work in cities rather than in small towns or in the countryside tend to have more options to consider, although correspondingly competition might be fiercer. And within local labour markets – whether defined as a city, a rural hinterland or a region – there are distinct submarkets, open to people with particular characteristics or skills. As geographer Ron Martin (2001: 461) argued, a local labour market is:

> A complex assemblage of segmented submarkets, each having its own geographies, its own employment and wage processes [subject to national regulation and the minimum wage in the UK] ... an assemblage of non-competing submarkets which, nevertheless, are linked together to varying extents via direct and indirect webs of local economic dependency.

These segmented submarkets are defined not only by the intersection of the demand for workers with particular skill levels and the availability of different types of job, but also by the ability or willingness of potential employees to travel to work. As the study of a declining labour market in Massachusetts by Susan Hanson and Geraldine Pratt (1995) showed, women are less likely than men to travel considerable distances to work as it makes the combination of employment and domestic responsibilities difficult. But this decision to search for work locally restricts women's options and deepens gender segmentation.

The geography of employment – the spatial distribution of different forms of work – affects people's life chances as well as influencing the form of social relations that develop in that area. This emphasis on the local nature of labour markets is one of the distinctive contributions that geographers have made to the analysis of work and employment. Here US geographers Michael Storper and Dick Walker (1989: 47) explain the significance of the local focus:

> Local labour markets deserve special emphasis because of labour's relative day-by-day immobility which gives an irreducible role to place-bound homes and communities.... It takes time and spatial propinquity for the central institutions of daily life – family, church, clubs, schools, sports teams, union locals, etc. – to take shape.... Once established, these outlive individual participants to benefit, and be sustained by, generations of workers. The result is a fabric of distinctive, lasting local communities and cultures woven into the landscape of labour.

This sense of place, based in a deep rootedness in the locality in which customs and local institutions are structured by the dominant form of work, has been a key feature of twentieth-century economic development in many industrial nations. In recent years, however, the significance of distinctive place-based local labour markets has declined, as new forms of work and the rise of mobile workers, as well as new connections across space as globalized supply chains and transnational corporations become significant, have affected the spatial scale at which workers are assembled. Its decline also reflects the rise of the service economy. Compared to manufacturing economies, regional and local differences in labour market participation rates, as well as in types of jobs, are less significant in service-dominated economies such as the UK, as Walby (1986) recognized more than twenty years ago. In Britain, women's participation rates, as well as the sorts of jobs they undertake, are regionally less distinctive than patterns of male employment participation during the Fordist period. In service-based economies, in which a high percentage of women are employed, the distribution of employment, especially those consumer service jobs catering for immediate demands, tends to be related to the distribution of the population. Thus, in most British towns and cities there is a similar range of employment opportunities in private consumer services such as hospitality, the retail sector, and leisure industries and in public services – in the health service, education and in local government: their size is related to the size of the local population. Well-paid and high-status producer service occupations in, for example, high-tech industries and in finance, business and legal services are more unevenly distributed. In these latter occupations, the southeast of England, especially Greater London, accounts for a high proportion of the jobs in these sectors (Allen 1992). This pattern is reflected in geographical variations in income and wealth (Dorling et al. 2007). At the scale of the region, the southeast is the most affluent, reflecting the dominance of the City of London in the British economy (Hamnett 2003; Massey 2007). Those regions that were previously reliant on older manufacturing industries, in the north of England, Wales and Scotland, are among the least affluent. In the recession that began at the end of 2008, however, in which the financial sector was deeply implicated, the southeast region found itself hit hard by rising unemployment.

The immediate locality, the residential neighbourhood, also has an impact on everyday lives, living standards and employment opportunities. Many services, such as schooling, are locally provided, with clear neighbourhood differences in standards and levels of achievement that affect children's aspirations and values as well as future labour market opportunities. As Bauder (2001, 2002) has argued, neighbourhood institutions play a powerful role in inculcating sets of cultural values. In British cities there are marked spatial variations in levels of income and wealth. Even in the most affluent British cities, in inner areas and outer local authority estates, there are concentrations of poor households. Parts of inner London, for example, especially in the east, include some of the poorest districts in the country as a whole; households there are increasingly reliant on poorly paid 'high-touch' jobs (Wills 2004; May et al. 2007) and are among the rising numbers of the 'working poor'. Elsewhere in the city, in Kensington and Chelsea and in Westminster, for example, some of the richest households in the UK live largely separate lives. Where their lives do come into contact, however, is in leisure spaces, where the working poor wait on the rich, and increasingly frequently in the home as the children of the rich are cared for and dirt is vanquished by the labours of the poor. One of the contradictions of the rise of service employment, explored in more detail in chapter 4, is that despite increasing spatial separation between the rich and the poor in many arenas of daily life and in residential segregation (Dorling et al. 2007), in growing numbers of homes they live/work in close contact.

Thus, as more and more people are drawn into waged labour, their connections to each other, to the organizations in which they labour, to the locality and to class segregation are changing. Employment, which was once a local affair in which people tended to be employed in the immediate vicinity in which they lived, often in locally owned firms, now links people across increasingly extended spaces, regions and nations, sometimes involving physical movement across space of both labour and capital, but also linking workers in particular locations into new networks of ownership. Changes to British tax laws and a relatively unregulated labour market have made the UK an attractive place for foreign investment, as well as a country where the penalties for making workers redundant or for disinvesting are less onerous than in many parts of the western world. The working class in the UK as well as the new global working class is, therefore, increasingly complex and diverse (Panitch and Leys 2000; Mason 2007). In the UK as a whole, about 10 per cent of the population was born elsewhere. In London this figure rises to 26 per cent of the population and to 30 per cent of all employed people: migrants are younger and so more likely to be employed than British-born inhabitants (Spence 2005). The countries from which these migrant workers have arrived have become increasingly diverse over the last two decades (Vertovec 2006), leading to enormous variations in, for example, skin colour, skills, national attitudes and family relations which are the basis for the establishment of hierarchies of eligibility within the potential labour force. As embodied attributes become a more significant factor in employment relations, this diversity and the consequent paradox of an increasingly globalized labour force performing place-tied servicing work is a key part of the explanation of how segregated labour markets work.

These flows of capital and labour across an expanding spatial scale have altered the links between waged work, nationality and locality, reforging the spatial division of labour, and increasingly bringing workers born in one nation-state into physical contact with those born elsewhere through new migration patterns, as well as connecting growing numbers of workers through ownership patterns. Labour, as well as capital, has become more mobile. Thus, in the core economies of the 'old' industrial west, there is a growing reliance on migrant labour from the third world to run key urban services and provide for all those tasks once undertaken in the home (Anderson 2001; Sassen 2001; May et al. 2007) as women enter the labour market in growing numbers. Migrants also work in sweated conditions in what remains of basic manufacturing industries such as textiles, clothing and electronics in first world cities. Parts of these same industries, however, have relocated to the border regions in third world economies, to export-processing zones in Southeast Asia or the maquiladoras of the US/Mexico border, for example, where labour costs are lower (Ong 1987; Oishi 2005; Wright 2006). Thus, what might be termed the old working class of the first world and the new working class of the third world increasingly are spatially contiguous in western metropolises, but also spatially differentiated by the dispersal of workers in a particular sector, or as employees of a single multinational company, across the spaces of national economies, raising new questions for managers and for labour organizers.

I explore these new spatial divisions of labour in a range of different forms of embodied interactive work in many of the case-study chapters, showing how, for example, sex work (chapter 5) and care work (chapters 4 and 7) in a global city such as London now employ workers from a wide range of countries outside the UK. In Chicago – the location for an exploration of the world of boxing (chapter 6) – new patterns of migration into the USA are changing the association between men of African origins with boxing. I also show how the geographic reach of contemporary capitalist organizations as well as the national origins of service workers has expanded. In all sorts of jobs, even the most mundane tasks of keeping the bodies of children and the elderly clean and safe, managers, workers and organizers have to cope with cultural and linguistic diversity, whether in negotiating agreement and compliance, or in organizing or defusing resistance. Significant social, local and national differences in customs, beliefs and cultures among a workforce that is increasingly diverse, as women, children, rural-to-urban migrants, ethnic minorities, refugees, asylum seekers and economic migrants enter labour markets previously dominated, in the west at least, by men means that 'cultural' understanding and connections have a growing salience in 'economic' organization (McDowell 1997; Du Gay and Pryke 2002; Jones 2008). Divisions of labour now cross, or are negotiated over, diverse and multiple cultural and linguistic spaces as well as geographic space or distance and so new ways of drawing in and constructing co-workers and of managing cultural differences among them are important in multinational spaces. Thus, 'globalization' is both affecting and is affected by social and cultural processes, producing both new forms of waged work and new senses of self-identity.

Structure of the Book

Theorizing the effects of this growing complexity is a significant challenge for labour market analysts. In chapter 3 I look at some ways to think about complexity as well as providing a guide to theoretical debates about corporeality, embodiment, emotion and affect, and sexuality and desire at work, with the aim of bringing together debates about different forms of labour market differentiation. While the work on embodiment provides an important way of thinking about service sector employment change, distinctions based on categorical and group differences – class, ethnicity and gender – remain a key part of understanding who works where in service economies. Before turning to these theoretical debates about difference, however, in chapter 2 I look in more detail at the current division of labour in the UK, addressing the growth of interactive embodied forms of work, as well as the social and spatial division of labour and the ways in which deepening patterns of inequality are emerging in a service economy. I also provide some statistical information on the current structure of the UK labour force. The emphasis is on the UK, but as an exemplar of trends that are emerging more widely in economies once dominated by manufacturing production but now increasingly reliant on services: economies that, as US lawyer Katherine Stone (2004) notes, have shifted from 'widgets to digits'. A pattern of deepening income inequality, as well as the rise of work that is insecure, precarious and impermanent, is evident in, for example, the USA and Canada, as well as in the UK and other European countries (Vosko 2001; McGovern, Smeaton and Hill 2004; Green 2006).

The rest of the book – chapters 4 to 8 – is about different forms of work and working lives, looking at who does what type of work, where it takes place and what tasks are involved in different sorts of jobs. In these chapters, I draw not only on my own previous empirical research in and about bars, cafés, hospitals and hotels, but also on numerous other fascinating studies that document social relations in different types of workplaces. There is now an enormous range of work that explores the ways in which the sex typing of occupations has been established and maintained over time (Bradley 1989) and the ways in which feminized jobs depend on the manipulation of stereotypical gendered attributes, especially women's bodies, emotions and their sexuality (Adkins 1995; Gherardi 1995). These arguments have been explored in case studies of, among others, air stewards (Hochschild 1983; Tyler and Abbott 1998), bankers (Halford, Savage and Witz 1997; McDowell 1997), beauty therapists (Sharma and Black 2001; Ahmed 2004), doctors (Pringle 1998; Moreira 2004), hairdressers (Gimblin 1996; Furman 1997), nurses and carers (James 1989; Diamond 1992; Lopez 2006; Daiski and Richards 2007), secretaries (Pringle 1988), sex workers (Chapkis 1997; Brewis and Linstead 2000a, 2000b, 2000c), shop assistants (du Gay 1996) and waiters and waitresses (Hall 1993b; Crang 1994; Fine 1998) in which men, as well as women, feature both as clients and as service providers, as owners, workers and employees. These studies reveal the multiple ways in which the commodification of gendered, racialized and classed bodies is produced and reproduced through workplace interactions and cultures in particular ways at different times and in different locations, establishing a new hierarchical structure of inequality in which embodied interactive work typically is poorly valued, underpaid and low status. These sorts of employment are the focus here, even though many types of high-status employment also involve 'body work' – by doctors, for example, on the bodies of others or by the investment bankers I have written about elsewhere (McDowell 1997) on their own bodies to produce a pleasing appearance that chimes with clients' views of an educated and skilled provider of technical financial advice.

In these chapters I have tried wherever possible to include the voices of the workers themselves, talking about their occupations and what the job means to them. I have also where it seems appropriate asked questions about how the research was undertaken, although I have not done this in every case as it would become too repetitive. In one of the chapters on masculinity at work (chapter 6) the methodological issues involved in undertaking workplace ethnographies are a central part of the argument. Reading about the different ways in which work is organized has been a great pleasure over the years. Most of us have a rather limited experience of different forms of work – we might know something about the jobs that our parents and other relatives do; we may have worked as students on a casual basis in some of the jobs that appear in these chapters – in a bar, as a room attendant in a hotel or a domestic assistant in a care home – but most of us know relatively little about the huge variety of service sector jobs and occupations that currently make up the division of labour in Britain. I hope that this book will partially correct this lack of knowledge, but better still that it will encourage you to undertake your own research into working lives in the service sector.

Part I
Locating Service Work

so I manuf
a sorvic - tam oco.

2

The Rise of the Service Economy

Nowadays most jobs call for a capacity to deal with people rather than things, far more interpersonal skills and fewer mechanical skills
Arlie Hochschild, *The Managed Heart*, 1983

This chapter and the following one provide the basic definitions of the concepts which reappear in the rest of the book. They build on chapter 1 where I briefly discussed the rise of the service economy and the reasons why questions about embodiment and emotional labour must be a part of any explanation of labour market segmentation, the construction of social identity and social interactions in the workplace in service economies. In chapter 2 I look in more depth at the consequences of the decline of manufacturing employment and the shift to a service-dominated economy in the UK, and the associated rise of a more diverse and unequal workforce. I assess optimistic arguments about the quality of working life in knowledge-based economies that were introduced briefly in chapter 1, as well as pessimistic debates about growing polarization in service-based economies. In chapter 3 I explore ways of theorizing the complexity of embodied identities in the labour market, examining how class, gender and ethnicity intersect in the allocation of service workers to different types of employment.

This chapter does three things. First, it addresses the basic distinction between work and employment, itself based on a set of assumptions about the place of men and women which is written onto their bodies. This discussion sets out the basic arguments about how certain types of work – typically, those undertaken by women – are associated with natural attributes of femininity and as a consequence devalued in the labour market. This is the key argument behind the book's claim for continuity compared to those theorists who have identified an epochal change and the rise of a 'new' economy, as well as those who argue that waged employment now plays a less central role in the social construction of identities than in the era of industrial capitalism and manufacturing dominance. Women's entry into the social relations

of waged labour but in the same types of work as they previously undertook
at home paradoxically represents both change and continuity. Secondly,
I define services: distinguishing producer and consumer services. Embodied
labourers work, in the main, in the consumer service sector, doing interactive work that often involves working on the bodies of their clients or consumers. I then establish the numerical significance of embodied interactive
employment in the UK. Thirdly, I describe the emerging social divide
between high-tech and high-touch work: a distinction that maps onto the
producer and consumer services divide as well as, in large part, gender divisions of labour. High touch work is the focus of this book.

What is Work?

Work in the widest sense consists of all those activities that are central to
material existence, to our place in the world: indeed, to all aspects of human
life. It provides us with sustenance, goods for exchange and, in most societies,
income. Work is what makes life possible, it is a way of producing things
to eat, to wear, to shelter in and to sustain not only ourselves but also all
those people who are considered too young, too old, too weak or incapable
of working for themselves. Work involves human effort – the application of
labour power – in order to transform material resources into goods for the
use of individual workers, their households and other dependents and for
exchange with others. Work covers all types of activities and there is a fuzzy
boundary between what is defined as work and what is regarded as play or
leisure activities. Fishing, for example, might be an essential activity in the
survival of some people and a weekend leisure activity among others. Similarly
conversation, song, home decoration and sexual relations are variously
defined as work or leisure/pleasure depending on whether cash is exchanged
between participants.

Work is a complex activity that takes place in a range of locations under
a variety of conditions. It sometimes, although not always, is rewarded by
payment in cash or in kind. There is a further set of costs and benefits associated with work, depending on its nature. Some types of work are hard,
dirty or dangerous; others take place in clean and comfortable surroundings
and, contrary to what might seem just, the latter are usually better paid.
Work also provides different degrees of status and respect, as well as access
to particular forms of association with colleagues and in the wider society.
It is part of the construction of social identity, of a sense of self-worth, and
brings with it not only sociality but social evaluation, as some forms of work
are regarded as more legitimate and more valuable than others. Further, as
feminist theorists have argued, the gender of the worker is associated both
with the financial reward and the perceived social status of the tasks in

particular job categories, as well as with the actual definition of work itself. The work that women do, as I explore at greater length later, typically is less highly regarded as well as less well remunerated.

Waged work also has to be regulated and controlled: workers typically have job contracts but these are often imprecise. Contracts may specify hours and pay rates, but how much effort is required and exactly what is the successful completion of tasks is usually less clear. As a consequence employers have to actively manage the labour process. Here too status distinctions are clear: well-paid and interesting work is often less highly regulated, leaving workers with a considerable degree of autonomy and personal discretion; other forms of work are more highly regulated and often socially alienating, as Marx argued about early factory employment. In service economies, the equivalent of the assembly line in a car factory (Cavendish 1982; Beynon 1984) is perhaps the fast food outlet (Gabriel 1988; Leidner 1993), where workers have to reproduce a tightly scripted performance under management surveillance, or the call centre, where speed is essential and workers are heavily monitored (Bain and Taylor 2000). To counter management control, workers have formed collective associations to represent their own interests, through professional associations and trades unions in the modern era. In the service economy, however, trade union membership has declined from its peak in the late 1960s. In 2006, about one in four workers in the UK were union members (ONS 2007), although there is considerable sectoral and regional variation: 40 per cent of workers in the northeast, for example, are members, reflecting this region's old industrial heritage, whereas in Greater London the proportion is about half of this.

There has been a long tendency in geographies and sociologies of work in capitalist economies to assume that waged labour is synonymous with work, that work, in other words, consists solely of waged employment and other forms of production are ignored. But work is not only undertaken in markets for financial reward; it takes place in a range of locations under different conditions of exchange (Tilly and Tilly 1998). The world of formal labour markets, jobs and occupations that is now seen as typical in the west developed unevenly over time and space and is still less dominant in many societies where informal labour markets are significant. Furthermore, the formal relations of waged labour encompass a variety of different ways both of organizing the production process and the relationship between employers and their workers, from multinational firms with international recruitment policies to family businesses and local profit-sharing worker cooperatives.

One of the consequences of the definition of work as waged labour, however, is that it excludes all that work that is undertaken in the home for 'love' rather than money. Millions of women, as well as many men, labour within their homes to ensure the social reproduction of their household. Meals get made, cleaning and childcare are undertaken, a wide range of products may be made at home – cakes, jam, clothes – and household members are also the recipients of a range of services including care, comfort, counselling, sex and entertainment. These services are freely provided, in the sense that they are not based on a monetary exchange, other than shared financial support within the household, often based on the earnings of one or more 'breadwinners' (typically a man until relatively recently). In the development of industrial capitalism many of the previously home-produced goods were commoditized, available for purchase in the market. It is only when this transfer occurs, when domestic services become part of the cash nexus, that those doing the work are counted as employed and so classified as part of the labour force (Waring 1988). Throughout the twentieth century, the range of previously home-based services also available as market commodities expanded.

Even after their transfer, however, the provision of 'household' services tends to remain feminized. The majority of workers are women and, in common with other female-dominated sectors of the economy – clerical work, primary school teaching and nursing – these jobs are low status and typically poorly paid. As I explore below, this shift of domestic services into the market has been part of the wider transformation of advanced economies from ones dominated by manufacturing work undertaken by men to service-based economies, in which both men and women labour for financial reward. In the new post-millennial economy in the UK, more than two thirds of all workers, over half of them women, are employed in the service economy. Women's participation has risen from one third to half of all waged workers over the second half of the twentieth century: in the main in public and private consumer services, which I define below. Like work undertaken in the home, for the major analysts of the changes in ways of working brought about by industrialization, service sector jobs were ignored or not defined as employment. For Adam Smith, the author of the *Wealth of Nations* (1776), work was not proper work unless the labourer was making things: material products embodying physical effort.

> The labour of the menial servant ... does not fix or realize itself in any particular subject or vendible commodity. His [*sic*] services generally perish in the instance of their performance, and seldom leave any value or trace behind them for which an equal quantity of service could afterwards be procured. (Smith 1986: 430)

Almost a century later, Marx argued along the same lines:

> Types of work that are consumed as services and not in products ... separable from the worker, and not capable of existing as commodities independently of

him ... are of microscopic significance when compared with the mass of capitalist production. They may be entirely neglected. (Marx [1868] 1976: 1044)

At this time Britain was emerging as the workshop of the world and was the major producer of goods. Indeed, until 1955 more than half of all employed workers were engaged in making things in the manufacturing sector. Nevertheless, a significant proportion of the population, even in the eighteenth and nineteenth centuries, worked in the service sector, but as many (although not all) of them were women, often working in other people's homes, their labour was ignored. As a consequence, it is almost impossible to provide an accurate assessment of how significant these forms of employment were, let alone who was employed in 'servicing' work and whether and how much they were paid. One of the interesting aspects of the growth of service employment as a proportion of the national labour force is the continuing significance of the labour of 'the menial servant' identified, albeit then ignored, by Adam Smith. In the 'affluent economy' identified by Green (2006: 6), where a growing proportion of the population are working for wages and so are short of 'free' time, there has been 'a return of servant occupations, there to pack bags, clean floors, secure property – a renaissance which mocks earlier expectations that the servant class had disappeared for ever in the first part of the twentieth century'. And as well as this class labouring in the homes of their employees, there has been a significant expansion of the same types of work undertaken in the public arena – in shops, offices, small firms and fast food outlets. Thus many of the women who serviced individual and familiar consumers in their homes, now provide similar services for large numbers of anonymous others in the marketplace (Glenn 1992).

The Expansion of Service Employment

As I argued in chapter 1, service employment now dominates advanced industrial economies, at least in terms of numbers of employees. Exchanging ideas and providing services has replaced the manufacture of objects. In the UK, for example, at the turn of the twenty-first century, 60 per cent of all men and 82 per cent of women in employment worked in the service sector, a higher proportion of workers than ever worked in the manufacturing sector. Twenty years earlier one in every three men in employment worked in the manufacturing sector. By 2001, the figure had fallen to one in every five men in employment; for women, the change was from one in five in 1981 to one in ten in 2001. This change was paralleled by a change in the distribution of jobs between men and women. In 1981 men held 3.2 million more jobs than women. By 2007 the distribution was almost equal: 13.6 men and 13.5 million women were in employment, although half of women employees worked on a part-time basis (ONS 2008a).

The service economy is also characterized by greater heterogeneity than the manufacturing economy in terms both of skill requirements and patterns of work across the day and the year. Both highly skilled and unskilled activities are expanding as some jobs are replaced by information and communication technologies but others demand personal service at the point of delivery. One in ten of all jobs were in financial and business services in 1981; the figure had risen to one in five by 2001, seeming to support claims about the rise of a knowledge economy. However, as Goos and Manning (2003) have shown in their assessment of the labour market in the last decade of the twentieth century, there had also been marked growth in 'bottom end' service sector jobs, which increased from 1.2 to 1.9 million (and remember this figure undoubtedly excludes a lot of poorly paid, often illegal, 'off the books' work) in Great Britain in the 1990s. The jobs that expanded most quickly in that decade included sales assistants, checkout operators, cooks, waiters, bar staff, youth workers, telephone sales, and security guards, as well as some growth in the numbers of nurses, hospital ward assistants and care assistants. Many of these jobs involve serving the needs of customers or caring directly for their bodily needs. The services on offer must be delivered as they are generated. They cannot be stored and used later. Consequently, the demand for flexible work, for services available across 24 hours, has grown, also increasing the diversity in terms and conditions of employment.

 Before estimating the extent of interactive service employment, I want to focus first on the question of whether this shift in the nature of employment means that the UK is now a *service* economy. I want to define services and distinguish between producer and consumer services as well as examining the continuities and changes in how services are provided. It seems self-evident that the key basis for the claim that there has been a shift to a service-based economy in the UK lies in the number of workers now employed in this sector. In other western economies there has also been a rapid and remarkable growth in the number of service workers. The service sector now provides employment for about three in every four waged workers in Western Europe, North America and Australasia. In 2001, for example, OECD data showed employment figures for the US, UK, Sweden and France as 75.2 per cent, 73.4 per cent, 73.8 per cent and 72.2 per cent, respectively, leaving the manufacturing sector as the place of work for something less than a quarter of all workers in these countries. Fifty years earlier about half of all waged workers were in the manufacturing sector. Its former dominance has disappeared, as debates in the 1980s about the long industrial decline in these countries make clear (Bluestone and Harrison 1982; Massey and Meegan 1982; Martin and Rowthorne 1986; Hutton 1996; Dicken 2003).

As Bryson and Daniels (2007) argue, however, this shift in the structure of employment does not necessarily mean that the UK has become a service economy. They insist that 'the shift towards service employment should not be equated with the demise of manufacturing nor the complete displacement of direct production with service work' (p. 7). They make this claim on the basis of a distinction between changes in employment and the rise of productivity in manufacturing which means that the value of manufacturing to the national economy has not declined at the same rate as employment. In 2004 in the UK, despite only 3.3 million people (16 per cent of total employment) working in the manufacturing sector, manufacturing goods still accounted for 55 per cent of total exports (Mahajan 2005). These earnings are five times higher than earnings from the export of all knowledge-based services (Coutts, Glyn and Rowthorn 2007). Thus, the income earned from the export of goods remains a significant part of the British economy as productivity gains and the shift into high-value goods is facilitated by technological innovations.

Producer services and the financial sector as the leading edge of growth (until the credit crunch)

Despite Bryson and Daniels' insistence that employment in services should not be equated with their economic dominance, there seems no doubt that Britain is a service economy. The overall value of manufacturing products in British exports continues to decline as many parts of the rapidly industrializing world – China, India, Southeast Asia – are shifting from low-cost mass produced goods into the kinds of specialist high-tech manufacturing, research and development where the west hoped to build comparative advantages (Dicken 2007). As a consequence, the value of producer and finance and business services in export earnings seems set to continue to rise in the west, despite the crisis in the financial sector that developed in late 2007. Until this financial crisis, still deepening in 2008, the financial services sector lay at the centre of economic and social theorizing about the 'new' knowledge-based economy. Financial and business services are distinguishable from the type of consumer services at the heart of this book. They are producer services which typically are the input to a further stage in the production process, and essential to the operation of the economy as a whole. The money markets and the range of business and legal services that support them are the quintessential example here. As a range of economists and geographers have argued until recently, financial services are (or have been) the leading edge of the British economy and essential to its global competitiveness (Allen 1992; Leyshon and Thrift 1997; Green 2006; Massey 2007). Indeed, in the 1980s and 1990s, financial services were identified as the key motor of an emerging global economy, characterized by new spatial patterns of increasing connections at an ever-larger spatial scale. Technological change, the rise of ICTs (information and communication technologies) and the apparently unparalleled and unstoppable growth of a knowledge-based economy seemed to indicate a radical break with the past. In 2008, as these claims started to unravel, it began to seem as if the economies of the western world may be on the edge of a second, and post-millennial, epochal change. However, it is too early to substantiate or dismiss this claim.

Manuel Castells (1996, 2001), an influential urban sociologist, argued that western economies had entered a new phase in their development in the 1970s based both on the development of new information technologies and telecommunications and on the internationalization of capital and its ownership. He captured the shift in the term 'network' societies, in which the leading edge of industrial development had shifted from the production of goods to the transfer of financial information, money and associated services. These new economies, he argued, are based on a *new spatial logic of accumulation* in which labour markets, financial exchanges and commodity markets have become increasingly interdependent and now operate at a global scale. The lynch pins of this global network economy are a small number of significant cities: those global cities where the corporate headquarters of multinational firms and the stock exchanges of the world are located (Sassen 2001). The three most significant global cities at the start of the new millennium were London, New York and Tokyo, where key corporate decisions are made and a huge volume of global financial transactions are processed. A second tier of global cities includes Frankfurt, Paris, Los Angeles and Singapore, as well as Beijing, Shanghai and Mumbai in the emerging global economies of China and India, respectively.

The quintessential workers of this new global network economy are bankers, global traders and – until the crash in 2000 – dot-com owners: the men (and they were mainly men) whom Tom Wolfe (1987) dubbed the 'masters of the universe' in his satirical novel *Bonfire of the Vanities* about bond traders in New York. In the UK, the City of London is at the heart of the knowledge economy. Under a decade of Labour governments – between 1997 and 2007 – the City prospered through deregulation of the banking industry and financial innovations, as growing numbers of non-UK owned firms were listed on the London Stock Exchange. Financial services in total account for almost 10 per cent of the gross national product, although the City itself is smaller. With Canary Wharf, it accounts for almost 3 per cent of GDP and 1 per cent of total employment. Despite these perhaps surprisingly small figures, its impact on the economy is much greater as it brings associated growth in a range of financial and business services and legal advice. The high salaries and huge bonuses of City workers fuelled rising house prices in the housing market of the southeast. For these high-status City workers, work seemed to have changed its nature, becoming flexible, de-standardized, detraditionalized and individualized (Beck 1992; Beck, Giddens and Lash 1994; Lash and Urry 1994; Bauman 1998; Sennett 1998). The old dominant pattern of nine-to-five employment for a single employer over the lifetime had disappeared, replaced by a more fragmented and plural employment system that is characterized by 'highly flexible, time-intensive, and spatially decentralized forms of deregulated paid labour' (Beck 2000: 77). This new system is based on networks rather than bureaucratic hierarchies in knowledge-based or informational economies in which highly skilled and individualized workers take risks and move between jobs, constructing mobile portfolio careers. The consequence of this new system is captured in Carnoy's (2000) definition of the new economy. He too emphasizes the significance of information technology in high-tech industries and the development of inter- and intra-firm networking, but he also identifies new ways of working and living that seem to justify the term 'new economy':

> It is a way of work and a way of life. Its core values are flexibility, innovation and risk. As the new economy becomes the main source of wealth creation worldwide, it infuses old industrial cultures with these values. It requires a workforce that is not only well-educated, but also ready to change jobs quickly and to take the risks associated with rapid change. (p. 1)

At the start of the 1990s, US economist Robert Reich (1991) identified symbolic manipulators and analysts as the key portfolio, risk-embracing workers in knowledge economies. Other scholars (Beck 1992; Bauman 1998) have explored the consequences of the new values identified by Carnoy, suggesting that the growing reflexivity or detraditionalization of these elite workers in high-status occupations increasingly frees them from the structural constraints of gender and class position, as well as from the constraints of bureaucratic forms of organization. Beck, for example, rather optimistically claimed that in these new economies 'men and women are released from traditional forms and ascribed roles' (Beck 1992: 105). Instead, elite workers apparently are able to construct an individualized workplace identity that is dependent on individual or team-based performance. *Performativity*, as much as credentials or skills, became recognized as a key aspect of personal identity in the arena and age of what Bauman (1998) termed aesthetized work. Image, appearance and the ability to convince have become essential parts of a service exchange. Thus, in knowledge-based economies, as the economist Francis Green (2006: 25) noted in the more non-committal language of his discipline, 'an important aspect of conventional economic growth theory has come to be centred on the role of human capital in delivering increasing returns and cumulatively higher growth rates'. Education and skills, but also style, presentation and the ability to convince, all matter in the marketing of producer services.

Like many theorists, governments in the western world had a similarly optimistic view of economic change, emphasizing the importance of creativity, intrinsic job satisfaction and greater freedom to influence the conditions under which waged work is undertaken. The British government, for example, argued that the newly emerging knowledge economy 'puts a premium on skills and knowledge at all levels but particularly on creativity and the ability to innovate' (DTI 2001). The European Commission (2001, 2002) in its development of the European Social Model emphasized both intrinsic job quality and the necessity of opportunities for life-long learning and career development as part of the goal of achieving higher quality in work, and member states all emphasized the significance of education as part of raising economic growth rates.

Much of the optimism of many of these accounts of new forms of work – and the new economy itself – began to dissipate when the dot-com bubble burst at the start of the new millennium. They were far more seriously challenged from 2007 onwards as economic analysts were forced by the pace of economic events to recall Marx's insistence of the inevitable tendency of capitalism to be marked by crisis. The financial sector in particular is characterized by temporal instability as profitable years and bull markets are succeeded by lower profits in bear markets. In 2007 and 2008 there were a series of financial crises that succeeded a decade primarily of financial growth, despite the dot-com crisis. The speed at which confidence in the banking system collapsed was astonishing, forcing unparalleled state intervention into financial markets in the USA and the UK, in part in response to profligate lending in risky housing markets. In 2007 the British government intervened to rescue Northern Rock and in 2008 the US government took over the two largest mortgage providers in that country – Freddie Mac and Fannie Mae – and a majority share in the insurance company AIG, but stood by and allowed the investment bank Lehman Brothers to collapse into insolvency. In October 2008, both the US and UK governments were forced to make available huge sums of money to the financial system to ensure its survival, fearing the consequences for the economy as a whole. Without this intervention, in the immortal words of the then-President, George W. Bush, referring to the US economy, it seemed clear that 'this sucker is going down'. The financial crisis is proving a serious challenge to the form of rampant, unregulated markets common in neoliberal economies and a salutary reminder that even in the most liberal of economies where there is an ideological belief in the deregulation of markets, state intervention remains crucial. The crisis was also a salutary challenge to the theoretical accounts of economic change that had dominated the social sciences literature during the years when the growth of the knowledge economy seemed invincible.

Consumer services and the role of the state in provision

A longer-standing challenge to claims about the emergence of a 'new' or knowledge economy and the rise of new ways of working lies in analyses of the second form of services – that is, consumer services. These services are, as the name implies, services that are dependent on local, face-to-face interactions between service providers and consumers. Co-presence is an essential characteristic of consumer service exchanges and the ratio between providers and clients is frequently high – sometimes one to one – making productivity increases almost impossible to achieve. The personal relationship cannot easily be reduced, replaced or reproduced through technological changes, although this is not impossible. A doctor, for example, might find it hard to see more than one patient at once, although the rise of online advice and NHS Direct in the UK – a call service offering emergency advice – is changing the interpersonal relationship that used to define a doctor-patient encounter. Caring for the elderly and the young is also dependent on a high ratio of carers to clients. Personal contacts cannot easily be replicated in other ways, although district nurses are now able to remind elderly patients to take their medication through the use of mobile phones rather than making a personal visit. However, it is clear that even these limited technological changes reduce the quality of the service. Within consumer services, a great deal of the quality and satisfaction in an employment exchange depends on the production of an empathetic emotional exchange by embodied workers, drawing on 'people skills' in close and often intimate encounters between workers and clients. All forms of interactive work consist of an interpersonal relationship between the provider and consumer of the service, that is place-specific, tied to the location where the exchange takes place. These are not the types of employment identified by the theorists of network societies that live on thin air, nor the knowledge sectors likely to provide high rates of economic growth and employment for a highly educated workforce.

Consumer services also differ from producer services in that they are far more likely to be provided in multiple ways. Consumer services are provided by the state as well as in the market, as the examples above suggest. A large part of health services provision in the UK, for example, is within the state sector, especially compared to provision in the USA where market-based healthcare, supported by various forms of insurance-based schemes, is dominant. Consumer services are also provided indirectly by not-for-profit or other private sector organizations acting on behalf of the state, as well as by not-for-profit organizations working beyond the state. And many of the services available in or beyond the market are still provided in a range of informal and unpaid services, typically by women. These are the services that ensure that daily life functions reasonably well for most people – what feminist theorists sometimes call the reproduction of everyday life (in its widest sense to include keeping people and houses clean, making sure people get off to work and school adequately fed and dressed, as well as having children).

It is ironic that some of the most influential commentators on the transformation of contemporary economies, including Castells, have largely ignored these forms of service provision, as Castells himself in the 1970s was a key exponent of the significance of the state in provision of 'collective consumption' services as essential for the functioning of capitalist societies. These are the services that spread the costs of daily and lifetime reproduction – of education, health, transport and in many societies, state housing – across the population as a whole rather than falling as a burden on individuals or employers. Despite variations in the level and form of state intervention between the corporatist societies of parts of Western Europe, the more generous collectivist societies of Northern Europe and the lean provision in the more individualist USA – what Esping Andersen (1990) characterized as the three worlds of welfare capitalism – the 'golden age' of twentieth-century economic growth and social insurance between the start of the twentieth century and the mid-1970s saw the rise of state spending on welfare services in their broadest sense. Furthermore, and what was not emphasized by Castells and many others, there was a steady growth of employment in the institutions of the welfare state in both health-related and social services as well as in parts of the transport system (classified as distributive services in UK official statistics). These institutions became particularly significant employers for women. In terms of the volume of waged work undertaken in the service economy at present, 42 per cent of the total (measured in hours) is in social services, compared to 22 per cent in producer services.

Castells (1977) argued that state intervention in the arena of social reproduction, welfare and transport was a necessary feature of advanced capitalist development, although his theoretical argument was overturned by the rise of Thatcher and Reagan and their commitment to cutting state expenditure. In a parallel argument, feminists asserted in the 1970s that women's unwaged household labour was also a necessary feature of capitalist production, but advanced industrial societies have demonstrated their ability to continue to function as many of the services supporting daily reproduction have moved into the market. Just as individual men showed an unexpected capacity to get by without clean shirts and hot meals, buying food and services including advice and sexual solace in the market (Ehrenreich 1984; McDowell 1991), so too did economies continue to function as workers no longer had access to state-provided housing, generous levels of income support or subsidised transport. It turned out that the view of capitalism that dominated analyses in the Fordist era was too optimistic: the capitalist system neither depends on women's unwaged work nor requires state-provided social services. The early twenty-first century capitalist system, especially in the UK and the USA, seems to be becoming more similar to a late nineteenth-century version than to the Fordist corporate capitalism of the middle decades of the last century (Panitch and Leys 2000; Harvey 2001). While the rolling back of the welfare state affected women's employment opportunities, nevertheless their participation rates continued to grow over the 1980s and 1990s and into the new millennium, exacerbating the decline of informal, community and home-based service provision and the expansion of market-based services. Public sector employment remains a central part of the UK labour market and is especially significant for women. Despite its decline over the 1990s, from 1998 it began to expand slightly and in 2005 provided employment for one in every five UK workers, the same proportion as the financial and business services sector.

Has interactive embodied work increased?

By definition only types of exchange that include co-presence in jobs where embodiment, emotions, personality and style are significant should be included in the numerical estimate of interactive body work. However, a wide range of occupations fall into this classification. Investment bankers providing advice in mergers and takeovers, for example, often meet their clients on a face-to-face basis and a good deal of business is enacted in circumstances that include social engagements where identities, personal style and empathy are crucial. More conventionally, however, embodied interactive work is seen as coincident with consumer services, social services such as health, welfare and education and all those types of work classified as personal services, including counselling, domestic services, the retail sector, hospitality, leisure and entertainment. In Great Britain at present, 16 per cent of the workforce is employed in social and personal services compared to 10 per cent twenty years ago and 7 per cent fifty years ago. The rise in the numbers of employees in interactive occupations in the second half of the twentieth century is shown in table 2.1, based on calculations by Nolan and Slater (2008).

I have used a broad definition of interactive embodied work here, including, for example, among professional occupations, teaching, as well as medicine, and among semi-skilled occupations, sales, bar work and catering. This encompasses almost 30 per cent of the current labour force. With the exception of doctors and dentists and elementary security jobs, the feminization of these interactive jobs is evident, as well as the noticeable expansion of social welfare occupations over the last half century.

Table 2.1 The growth of interactive service employment, Great Britain, 1951–2001, 000s

	1951			1981			2001		
Professional occupations	m	f	all	m	f	all	m	f	all
Medical and dental practitioners	52	8	60	74	22	97	106	63	170
Social welfare	18	15	32	61	76	137	79	206	285
Other medical	58	224	282	86	603	690	118	623	741
Teaching	135	215	350	323	437	760	330	645	975
Semi-skilled occupations									
Sales	532	648	1171	228	834	1062	467	1360	1827
Personal services; leisure	211	621	832	186	813	999	306	1549	1854
Cleaning and related	68	413	482	156	544	700	186	462	648
Bar work and catering	59	329	388	77	490	568	181	441	662
Elementary security	50	1	51	110	24	124	149	32	181

Source: Nolan and Slater's (2008) calculations from Great Britain Census.
Figures may not sum due to rounding.

Other analysts of 'body work' (Wolkowitz 2006; Cohen 2008) have adopted a more restrictive definition, including only those occupations that involve actual physical contact with and/or manipulation of others' bodies. This tighter definition reduces the proportion of body workers in the UK to about 10 per cent of all employees. Table 2.2 shows the numbers, significance and gender of the main categories of 'body workers' using this more restrictive definition.

There are several striking features revealed in table 2.2. The first is the significance of employment in a range of occupations in the health services: almost 60 per cent of all body workers are looking after the bodily needs of the population. The second noticeable feature is the predominance of women among body workers, especially in the lower-status occupations. In all the occupations except those that are associated with masculine strength and protectiveness (and, interestingly, dead bodies, although strictly speaking this should be excluded from my definition of 'interactive' work), men are the minority, although, as the table shows, when health-related jobs are divided

Table 2.2 Employment in 'body work' in Great Britain, 2005

	Number	% of all 'body' workers	% female
All employees	25,171,130	–	43.5
All body workers	2,932,679	–	72.3
Health service professionals (doctors, opticians, dentists)	223,994	7.6	37.9
Health associate professionals (nurses, paramedics etc)	613,011	20.9	82.6
Healthcare and related personal services (nursing auxilliaries, house parents, care assistants etc)	898,364	30.6	86.1
Childcare and related personal services	330,882	11.3	97.1
Therapists	139,739	4.8	84.9
Sports and fitness occupations	85,759	2.9	51.5
Hair and beauty salon managers	30,919	1.1	73.5
Hairdressers and related occupations	189,914	6.5	89.8
Personal service n.e.c. (undertakers and mortuary assists.)	16,173	0.6	36.6
Protective service officers (armed forces, police, prison officers, security managers etc)	68,498	2.3	11.6
Protective service workers (below officer rank)	335,426	11.4	19.1

Source: *Labour Force Survey*, Spring 2005, ONS

by skill level and social status men dominate here too. In all other forms of body work, the traditional associations between femininity, empathy and caring, and a 'natural' association with the messy business of bodily care and the management of its secretions and odours, codes these jobs as female and as low-wage, low-status forms of employment. The expansion of body work, its feminization and its immediacy – a sick, disabled or dependent body has to be cared for in situ and immediately, just as a hungry body has to be fed – is connected to the rise of precarious and 'flexible' work and to growing income inequalities in the UK labour market, as I explore in general in the next section and through case studies in chapters 4, 7 and 8.

Inequality, polarization and social divisions in the service sector

If the nature of officially defined employment changed over the last half century or so from the making of goods to the provision of services, so too has the distribution of the rewards of waged work between the social classes, between households and between individual men and women. As I have already argued, more and more of the population have become workers, producing an economy where paradoxically although wage work is more significant, its rewards have become less certain. Forms of precarious and non-standard attachment to the labour market have expanded in association with the rise of services. This is in large part a reflection of two features of consumer services already outlined: first, such services cannot be stored but demand immediate co-presence; secondly, as productivity gains are difficult to make, ways of cutting costs are important. This has resulted in greater 'flexibility' in the use of labour through part-time, casualized and temporary contracts (Smith 2005). Women have always worked on a 'flexible' basis: this is not new for them, but men are now experiencing greater uncertainty and 'flexibility' than in the Fordist era.

Overall, a more polarized labour force is emerging in the UK, with the middle range of jobs and incomes experiencing a relative decline (Green 2006). In the last two decades, the extent of inequality in earnings and in standards of living between those with degrees and 'good jobs' (the two are not always co-incident) and those with few credentials, in 'bad jobs' in low-status work, has become more pronounced (Goos and Manning 2003, 2007). This has resulted in a new division in society among the working-age population. Instead of (or as well as) a division between the more fortunate in waged work and the less fortunate who are unemployed, there is an emerging division between the majority of employees and those whom Polly Toynbee (2003) and Barbara Ehrenreich (2001) have termed the 'working poor' – workers whose wages are insufficient to permit a decent standard of living, often employed in casual and insecure forms of work. Many of the working poor are those who are employed in the sort of caring, embodied and emotional work that is the subject of this book.

Inequality in income, wealth and earnings in the UK as a whole, as well as the emerging polarization in job quality and employment conditions, reflects not only the growth of employment in consumer services but also the attitudes of the government towards acceptable degrees of inequality and the extent of income support through the benefits system, as well as changes in family structure, the nature of the job market at different times, unemployment rates and the age of the population, as pensioners are on average poorer than the working population. Inequality in the income distribution widened significantly under the Thatcher-led Conservative administrations between 1979 and 1991 and then remained largely static from 1992 onwards. During the 1980s the whole wage distribution widened as successively higher percentiles of the wage distribution experienced higher relative wage growth (Machin 1999): in other words, the rich got richer. In the 1990s, the difference between the wages of workers at the bottom end of the income distribution and those on average wages began to narrow somewhat, although inequality continued to expand when the top decile and average wages were compared (Machin 2008). So in this decade and into the 2000s, although the rich continued to get richer, the position of the poorest was more stable. In 2005 the richest 10 per cent earned nine times the income of the poorest and this difference is greater if wealth as well as income is compared. Throughout the first years of the new millennium, unemployment rates in the UK were relatively low at around 5 per cent of the working-age population. Growing numbers of people in employment, the minimum wage and tax credits that supplemented the incomes of the poorest made a difference to those at the bottom end of the income distribution. An Organization for Economic Cooperation and Development study (OECD 2008) found that income inequality began to decline in the first five years of the new millennium, although the earnings gap between the rich and the poor was still 20 per cent wider in 2005 than in 1985. The UK remains one of the most unequal societies in Europe, with low rates of social mobility. From 2008, in large part as a consequence of the financial crisis, unemployment rates began to rise. As the cost of living was also rising, the position of the poorest households may once again deteriorate as the rising cost of basics such as food and fuel hits the poor hardest.

Table 2.3 shows the proportion of the total population and the working-age population living on less than half the median income (a commonly used poverty line) between 1979 and 2007. The same trend is evident among both groups and this table shows that the active labour market policies of the Labour governments have had relatively little impact at the bottom end of the distribution. Although the introduction of the minimum wage in 1999 initially improved the relative position of the poorest paid workers, as only 6–7 per cent of all workers were included in those eligible (many of them women working on a part-time basis), the effect was not highly significant (Dickens and Manning 2002). The 9 per cent of the population who lived on less than half the median income in 2007 includes 3.4 million working-age individuals (5.3 million after housing costs are taken off total incomes). Women, single parents and Black and minority ethnic (BME) people, especially people of Bangladeshi and Pakistani origins, are over-represented among the low-income population, as are people relying on part-time work or who are self-employed. The growth of women's participation rates has, however, improved the position of households which include a 'working' couple, especially if each worker is employed full time, but also when one, and typically a woman, is working on a part-time basis.

An analysis of the longitudinal British Household Panel Survey (BHPS) (Jenkins and Rigg 2001; Jenkins and Cappellari 2004) has revealed that the individuals with low incomes are a dynamic rather than a static group, with substantial (although short-range) income mobility from one year to the next. In other words, there is movement into and out of poverty, rather than the poor always being poor. About half of the group with below half median

Table 2.3 Percentage of the population falling below 50 per cent of the contemporary median income in the UK, 1979–2007

	All individuals	Working-age adults
1979	5	4
1981	5	4
1987	8	6
1988/89	11	9
1990/91	13	10
1991/92	12	10
1992/93	12	10
1993/95	11	9
1994/96	10	8
1995/97	10	9
1998/99	11	8
1999/00	10	9
2000/01	10	9
2001/02	10	9
2002/03	10	9
2003/04	10	9
2004/05	10	9
2005/06	10	9
2006/07	11	9

Source: Adams et al (2008) p. 40 and 115; figures between 1979–1997 from the Family Expenditure Survey (single years initially and then combined two calendar years); later figures from the Family Resources Survey (for single financial years).

income in one year were not poor in the next. But as they moved up, others just above this line fell below it. Over a six-year period, about a third of all individuals were poor at least once, although only 2 per cent were poor in all six years. This fluctuation reflects the increasing precariousness of low-status and low-waged employment at the bottom end of the labour market and the tendency of low-waged workers to experience a cycle of transitions between bottom-end jobs and unemployment (Stewart 2007).

Do the transformation theorists acknowledge the growth of 'poor work'?

Despite the focus on workers in the leading-edge sectors such as high finance, communication and the media, among the 'transformation of work' theorists, some adherents did recognize the growth of low-wage work, especially those forms that involve the servicing of the knowledge workers' daily needs (Castells 1996; Sassen 2001). In his thesis about the rise of a network society based on a new spatial logic of accumulation, Castells, for example, identified growing numbers of low-wage workers on whom this new economy was dependent. Beck, too, argued that paralleling the growth of detraditionalized workers there was an expanding group of less skilled workers whose fate was to become increasingly redundant or replaceable (Beck 2000). This growing division between workers is one of the key features of the emerging polarization. On one side of the divide is what Castells (2000) termed masculinized 'self-programmable' labour: workers with high-level skills and credentials in career positions that demand rational and cerebral skills and aptitudes and bring in their wake the prospect of prosperity, working primarily in producer services or professional occupations, such as medicine. On the other side is low-skilled, often uncredentialized 'generic labour' where workers of both sexes labour under 'feminized' conditions, with low levels of security and poor pay, servicing the needs of consumers.

There is a growing literature that documents the ways in which employment at the bottom end of the economy for these 'generic' workers has become increasingly precarious and insecure (Cully et al. 1999; Keep and Mayhew 1999; Ehrenreich 2001; Peck and Theodore 2001; Vosko 2001, 2006; Goos and Manning 2003; Fudge and Owens 2006; Green 2006). Despite the British government's emphasis on raising educational standards, many young people still complete compulsory schooling without gaining credentials and find it hard to gain access to permanent employment. In the UK in 2007, for example, almost 1.3 million young people aged between 16 and 24 were without work and not involved in education or training – a category known as NEET – and most of them left school as soon as they were legally allowed to do so (Prince's Trust 2007). They have few prospects for an affluent working life and seem doomed to be trapped in unskilled work at the bottom end of the labour market (McDowell 2003), as degree-level qualifications have become a requirement for a growing proportion of jobs. Between 1986 and 2001 the proportion of jobs requiring professional or degree qualifications rose from 20 per cent to 29 per cent in the UK (Felstead, Gallie and Green 2002). Nevertheless, as well as the high-paid jobs for which educational and professional credentials are necessary entry requirements, low-paying jobs have also been among the fastest growing over the same period. These include jobs such as security for businesses and in the leisure industry, retail employees, services for businesses, such as providing and arranging flowers, providing specialist catering or messenger services, telesales staff, hotel workers and care workers. As Green (2006: 36) suggests, 'such jobs are not those that spring to mind with the vision of the knowledge economy', nor do they bring with them high rewards or a high-quality working environment.

These workers – the low-paid 'generic' workers identified by Castells, often undertaking demanding but demeaning work – are trapped in the global economy, in Castells' terms working in the 'space of places' rather

than the 'space of flows', unlike the mobile middle classes who have greater
freedom to move across time and space (Urry 2000). And yet, as I shall
show, generic workers have also often been internationally mobile. At the
local scale, there is a growing divide between the more affluent and the
working poor reflected in patterns of spatial segregation. Affluent knowledge
workers employed in producer services often live in increasingly segregated
parts of towns and cities where these middle-class households, many with
two wage earners, use their growing wealth to separate themselves in space
from the new poor through gated communities, private transport and the
growing use of private rather than public sector services (Skeggs 2004a;
Dorling et al. 2007). Lisa Brush (1999) documented a similar polarization
and separation in the workplace as high-status workers seldom see the
'others' who service their work-spaces as they work different hours or are
actually ignored (Ehrenreich 2001; Toynbee 2003). She labelled the workers
at the high-status end of the new economy 'high-tech', whose affluence and
social status differentiates them from the 'high-touch' workers at the bottom
end, who provide care and do the dirty work necessary to keep the new
economy functioning and without which daily life would be impossible.
Naisbitt, Naisbitt and Philips (2001) used the same high-tech/high-touch
distinction in a book about new technology and its impact on daily life.
As Brush argued, the high-touch workers typify the types of embodied, emotional work that have increased so significantly in recent years.

    The high-tech/high-touch distinction, in large part, maps onto newly
emerging gender divisions in the service-dominated economy – both between
men and women and between women – recasting relationships between
gender, generation and occupational status. Although it has been argued
that emerging gender divisions are merely deepening older divisions in which
women are largely segregated into appropriately feminized occupations
(Adkins 2000), I suggest that the reshaping of older forms of both class and
gender divisions is evident (McDowell 2006a). The 'high-tech' and 'pro-
grammable' occupations in information-based production and services which,
as I noted above, require the possession of high-level educational and professional credentials, are now increasingly open, at least at the bottom rungs of
the career hierarchies, to women. Girls and young women are gaining academic credentials in growing numbers, now matching or even out-doing
young men's performance in, for example, school leaving certificates. Further,
young women are entering professional training courses in growing numbers, enabling them to challenge men's previous domination of high-status
occupations such as the law, medicine and finance (Crompton and Sanderson
1990; Crompton 1999). This success, however, opens up new class divisions
between workers, as both educated men and women enter the high-tech
world, but less well-educated women and men increasingly are trapped in
high-touch jobs: in retail, entertainment and commodified personal and

health care. Here women may have an advantage over the growing numbers of young men who, unable to find work in a shrinking manufacturing sector, are forced to turn to the service sector. These high-touch jobs draw on traditional feminized skills of empathy, care and servicing others, rather than attributes typically associated with masculinity.

As I explore in chapter 8, young men with low levels of educational attainment who have to find work in feminized 'serving' jobs are at a disadvantage compared to young women, often finding it hard to produce an appropriate deferential performance. This shift in employment patterns thus challenges traditional associations between masculinity and waged work established in the Fordist era and recuts older patterns of gender segregation in the labour market. As manufacturing employment disappears, and with it older patterns of stable work, at least for the male working class, men who leave school with few qualifications are no longer able to secure the sorts of relatively well-paid employment that they used to (Fine and Weiss 1998; Alcock et al. 2003; McDowell 2003). One of the indications of the changing associations between gender and low-wage service work is the increasing participation of men in low-wage jobs that were mainly the preserve of women until the 1990s. In the last decade, men have made inroads into many strongly female-dominated jobs (those with more than an 80 per cent female share), such as telephone operators, assistant nurses, retail cash and check-out operators, catering assistants and cleaners. In most cases, these trends continued into the new century, in some cases transforming work which was predominantly female in the earlier 1990s to predominantly male in 2005 (e.g. chefs/cooks and shelf fillers) (Grimshaw and Rubery 2007: 99).

In a pessimistic account of the nature of employment change, largely based on men's experiences, Richard Sennett (1998, 2006) has suggested that the unstable, fragmentary institutions of what he terms 'the new capitalism' have led to the 'corrosion of character', producing new forms of social and emotional trauma, as men in particular find the older certainties of the workplace dissolving. Elliott and Atkinson (1998) also argued that rising uncertainty is a feature of contemporary capitalist societies, despite (until 2008 when unemployment rates began to rise) almost two decades of growth in the UK. In the 1980s unemployment rates in the UK were almost 12 per cent, but at the start of the new millennium they had fallen to under 5 per cent. Rates began to rise again in 2008 (5.7 per cent in August of that year) and were predicted to increase as the effects of the financial crisis were felt throughout the economy. Despite high participation rates, Sennett argued that the world of work is much less certain than in previous eras. Employers, he suggested, no longer feel loyalty to their workers nor to the localities in which these workers live and labour. Increasingly, multinational firms, both in manufacturing and in the service economy, are able to change locations and recruit new labour forces in a search for lower costs or higher profits.

At the same time, the average job tenure for male workers has declined and so, Sennett suggests, workers' sense of their self-identity has shifted from the workplace to the locality and home-based interests have replaced work-centred networks. Sennett also argued that this shift was leading to defensive exclusion of 'Others' from locally based community networks, cementing patterns of residential segregation on the basis of class and ethnicity.

Sennett's arguments have been influential and undoubtedly capture some of the significant recent changes in the nature of work and workplace relations, as well as non-workplace formations of social identity. However, like many of the other 'transformation' theorists, he both exaggerates the changes and neglects to examine women's labour market experiences. For all workers, job insecurity (as measured by fear of job loss) declined between 1986 and 2001 (Green 2006: 133), although it is slightly higher for men than women at both dates. And for many women, job tenure has increased rather than decreased in the shift to a service-based economy, in association with overall rises in female labour market participation rates (Doogan 2001; Berthoud 2007). In part, this reflects women's greater concentration in public sector employment where job tenure is longer: 65 per cent of public sector workers are women, compared with 41 per cent in the private sector, and 57 per cent of all public sector workers in 2005 had held their current job for five years or more compared with 45 per cent in the private sector. Furthermore, women's average working hours rose between 2001 and 2005 (ONS 2005). Although women remain overwhelmingly concentrated in secretarial and administrative jobs, small numbers are moving into some of the previously male-dominated occupations, including personnel and training management, banking, the police service, educational officers and solicitors (Grimshaw and Rubery 2007) and divisions are opening up between well-educated highly paid women workers and others. Indeed, it might be argued that educated women are adopting the characteristics of male workers, whereas low-skilled and less well-educated men are increasingly constructed as feminized workers (McDowell 1991, 2001). In a service-dominated economy, as I shall show, the traditional advantages associated with masculinity have largely disappeared for men at the bottom end of the labour market.

Despite these shifts in status and identity, however, traditional patterns of gender segregation remain evident in the service economy, paralleled by a stubborn gender pay gap. Women in full-time employment in the UK earned in 2007 about 88p for every £1 that men in full time work earned. When women in part-time work are compared with men, the gap is yawning. These women earn 40p for every £1 for a man in full-time work. Women are still the majority of workers in particular sectors in the labour market – typically, the ones that are low paid – and gender continues to define occupations. Women hold only 34 per cent of managerial and senior professional occupations, 23 per cent of process, plant and machine operative jobs and 8 per cent

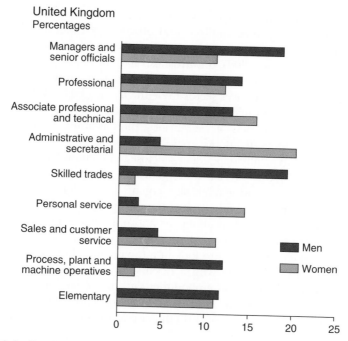

United Kingdom
Percentages

Figure 2.1 People aged 16 and over in employment: by sex and occupation, 2007, UK.
Source: Labour Force Survey, Office for National Statistics, 2008.

of jobs in the skilled trades. In personal service occupations, on the other hand, women are 84 per cent of the workers. Figure 2.1 shows how all men and women in employment are represented across the different occupational sectors, revealing the gender-specific patterns. Paralleling the pattern found in body work (tables 2.1 and 2.2), men are over-represented in the high-status group of managers and senior officials as well as in the skilled trades. And, although the proportion of women in administrative and secretarial jobs has fallen in recent years, this is still the job category in which they are most likely to find employment. Women are also over-represented in personal service and in sales and customer services, in jobs such as hairdressing, childcare and in the retail sector (ONS 2008b: 51), as the earlier tables revealed.

It seems clear, then, that the labour force in the twenty-first century is both more divided and increasingly diverse in its social attributes, as more women, more migrants and more workers of different nationalities enter into the social relations of waged work. However, it is also clear that most workers are in the main better educated than in previous generations and that many of them are using their skills in their working life, although it is important not to assume that skill levels equate with a high quality of working life. Some skilled workers, for example, may be trapped in jobs where their skills are under-utilized and others may experience relatively little autonomy or increasing stresses of deadlines. The financial crisis that developed from autumn 2007 onwards also had an unprecedented impact on the opportunities for highly educated City workers and challenged the claims of the financial services sector to be the leading edge of the economy and the motor of economic growth, at least in the short to medium term. As well as growing numbers of 'better' jobs, there is also a growing proportion of hard-to-automate, poorly paid bottom-end jobs in service-dominated economies, typically dependent on close interactions between workers and customers. In this diverse and polarized labour market, embodied attributes such as sex and gender, skin colour, age, weight and accent are increasingly important at both ends of the job spectrum, at the top as well as at the bottom end, as ideas about the appropriateness and suitability for different types of work are mapped onto the diverse characteristics of workers. In the next chapter I turn to look in more detail at how to theorize this growing complexity, starting with a discussion about the growing significance of sexuality, desire and embodiment in interactive forms of work in the service economy, before looking at the connections between the key social divisions of class, gender and ethnicity in interactive employment.

3

Thinking Through Embodiment: Explaining Interactive Service Employment

A key component of work performed by many workers has been the presentation of emotions that are specified and desired by their organizations.
J. Morris and D. Feldman, *Academy of Management Review* 21, 1996

In the second of this pair of definitional chapters, the focus shifts from the economy itself to conceptual issues about the nature of embodied work in service economies, looking at the ways in which theorists from different traditions have explained the coincidence of service work, embodied labours and the performance of social identity. In service economies, bodily labour – in the sense of an embodied performance – is a key part of many occupations, not only of the high-touch jobs explored here but also of some of the high-status, high-tech jobs that at first sight seem to depend entirely on cerebral, disembodied activities. As the quotation at the head of the chapter makes clear, however, waged labour in a wide range of occupations and professions in service economies increasingly depends on the manipulation of emotions to produce an embodied performance that meets the demands of employers. In both high-status and low-status work, bodies and emotions matter. However, there is still a clear status hierarchy that in part mirrors the older mental/manual labour division in manufacturing economies. In embodied interactive forms of work, in general the closer the contact is between the bodies of service providers and service purchasers, the lower the status of the work and, usually, the lower the financial reward.

From Labouring Bodies to Desirable/Desiring Bodies

As I established earlier, consumer service sector work almost always involves a direct relationship with a client, customer or consumer. It is work in which employees 'sell' themselves as part of the transaction (Hochschild 1983).

The bodily presentation and performance of employees are crucially significant in interactive service jobs or 'body work' (Wolkowitz 2002, 2006). Further, the jobs and occupations that are currently expanding fastest not only often demand a focus on the bodily performance of workers themselves, but also involve work on clients' bodies, including the 'adornment, pleasure, discipline and care of others' bodies' (Wolkowitz 2002: 497). In both aspects of body work – the production of an embodied performance by workers and the labour involved in the care of others' bodies – social relations based on sexuality, pleasure, desire and fantasy play an increasingly significant part in the employment relation: in hiring and firing, in acceptable workplace performances, in interactions between workers and customers or clients and in the ways in which social attributes are attributed to different types of work and which correspondingly are differentially rewarded. The growing significance of commoditized forms of body work has recast the associations between the social relations of production and the social construction of identity for increasing numbers of employees.

As Macdonald and Sirianni (1996: 4) argued:

> The assembly-line worker could openly hate his job, despise his supervisor and even dislike his co-workers, and while this might be an unpleasant state of affairs, if he completed his assigned tasks efficiently, his attitude was his own problem. For the service worker, inhabiting the job means, at the very least, pretending to like it, actually bringing his [sic] whole self into the job, liking it, and genuinely caring for the people with whom he interacts.

Macdonald and Sirianni rather hedge their bets here about the difference between a pretence of liking work and the emotions of genuine caring that might be involved and there has been a long debate about whether emotions at work are manipulated or authentic (Ashforth and Humphrey 1995; Mann 1999; Bolton and Boyd 2003; Erikson 2004; Sanders 2005). Hochschild (1983), in her original formulation of the notion of managed emotions, argued that emotional labour is used by organizations to control their workers, and so leads to alienation for individual employees (Leidner 1993; Pierce 1995). Other studies have found that workers find pleasure and satisfaction in drawing on their feelings to establish emotional connection between themselves and their customers or clients (Bolton and Boyd 2003; Erikson 2004). Nevertheless, the main point remains. Waged work increasingly demands and depends on an embodied performance in which emotional connections are a crucial part of the service exchange. The clothes, appearance, weight, facial expressions, gestures, sociability (whether 'real' or pretended) and the general bodily presentation of self that mark out an appropriate performance have grown in significance as essential attributes of acceptable service sector workers.

It is salutary to recall, however, that in his classic *White Collar Work*, first published in 1953, the US sociologist C. Wright Mills already recognized

the key features of what is now termed emotional, interactive or body work
and assumed by many to be a relatively new phenomenon in the contemporary workplaces of advanced economies. He argued thus:

> When white collar people get jobs, they not only sell their time and energy
> but their personalities as well. They sell by the week or the month their smiles
> and their kindly gestures, and they must practise the prompt repression of
> resentment and aggression. For these intimate traits are of commercial rele-
> vance and are required for the more efficient and profitable redistribution of
> goods and services. (Wright Mills 1953: xvii)

These intimate traits, Mills recognized, are shaped and managed not only by
workers themselves, but by managers, as white collar workers increasingly were
trained to produce acceptable performances and to shape their bodies and
personalities to benefit the 'corporation'. Mills' recognition of the importance
of emotional work more than half a century ago is further evidence challenging
the epochal transformation claims discussed in the previous chapter.

Gender, sexuality, emotions, performances and organizations

Until the late 1970s, however, C. Wright Mills' early analysis of white collar
'personality' work was largely ignored until a growing body of feminist-inspired
scholarship began to influence analyses of the labour market and organizational behaviour in sociology, business studies and economics. Ideas about
how gender shaped both individual workplace performances and the structure
of organizations, allied to insights about the significance of emotions at work,
produced innovative analyses of the emerging form of service sector organizations and employment. For many years, too many social scientists, especially
geographers, ignored the personal attributes of workers. Employees en masse
were seen as labour power which varied, for geographers by region or nation,
and which embodied a rather general notion of a 'tradition of skill'. Sociologists
and economists preferred the concept of 'human capital', which included educational and skill endowments. As individual actors, workers were seen as
rational economic 'men' seeking work that was as well remunerated as possible
in order to meet their daily needs.

Workplaces or organizations clearly differ in the terms and conditions
of employment that they offer, but like their employees, it was assumed
that they too were rationally organized in order to make maximum profits.
A significant challenge to the notion of rational bureaucratic organizations or
disembodied non-sexual (and by default masculine) workers that dominated
economic geography, sociology and other associated disciplines came in
work that might be grouped under the heading of 'gender and organizations'. Here Joan Acker (1990) challenged the notion that organizations
consist of profit-maximizing institutions that, through the employment of

hierarchies of employees without dependants, achieve market-defined ends. In this rational view of the world, employees, whose skills and knowledge best fit the goals of the organization, are appointed, rewarded and promoted on the basis of an objective evaluation of their performance. Instead, Acker argued, organizations are seldom rational or objective, but rather their structures, cultures and everyday practices are imbued with essentialist and non-essentialist assumptions about gender and sexuality which operate to consistently benefit certain workers, typically white, heterosexual men. The disadvantaged include a cast of 'Others': women, people of colour, less physically able workers and people with alternative sexual identities. These Others may be excluded or alternatively constructed as less suitable, inferior workers and restricted to a narrow range of jobs and occupations that are regarded as congruent with their gendered and sexualized identities. In consequence the division of labour per se and organizations and their practices are deeply gendered. Both conscious and unconscious practices produce and maintain gender inequality in the workplace.

Acker's work and the organizational case studies that followed it built on Rosabeth Moss Kanter's book *Men and Women of the Corporation* (1977). There she documented men's dominance of high-status occupations and positions, showing how men's power depended on women's support services, both in the workplace and the home. Thus, white (heterosexual) men, with minimal involvement in procreation and highly controlled emotions, reliant on someone else (usually a woman in the home) to undertake the messy and emotional tasks of daily reproduction, climbed the greasy pole to promotion in the workplace. A wide range of work in the last fifteen years or so has documented the ways in which this particular unsexed, independent bodily norm is part of the maintenance of gender inequalities in the workplace. Women are constructed as out of place in workplaces, especially those where the work is based on a version of rational, cerebral masculinity, but also where bodily strength is required. Their lack of conformity to a masculine bodily norm confines women to a narrow group of 'appropriate' occupations, as well as saturating the very definition of different tasks and jobs with gendered attributes. The various ways in which the discursive construction of organizational practices produce and maintain patterns of gendered and sexualized behaviour on the 'shop floor', in banks and laboratories, and in high-status management jobs, as well as in older forms of work based on brute strength and endurance (in mines and machine shops, in steel works and the army), have been documented in fascinating detail: some of these studies are the subject of later chapters. It is now recognized that:

> Not only are there gendered assumptions built into most job descriptions and job assignments, as well as variations in pay scales and occupational ladders/ promotion schemes, organizations also sexualize workers – presenting authority

and physical labour as testaments to heterosexual masculinity, and good looks, 'service with a smile' and covert sexiness as evidence of heterosexual femininity. These norms and expectations are maintained by open and hidden harassment and subtle and blunt sanctions by workers of each other and by bosses of workers under them. (Ferree, Lorber and Hess 1999: xxix)

If femininity structures less well-regarded jobs, masculinity is associated with management skills. Collison and Hearn (1994), for example, in empirical analyses of managers in several institutions, distinguished the varying ways in which male managers manage and regulate their co-workers through constructing alternative versions of masculine identities that construct or preclude dissent through coercive or cooperative strategies. Their strategies are variously based on versions of male authority or male trust and cooperation that differentially and unequally position male subordinates and peers but which also act to affirm masculine solidarity within organizations by excluding women and behaviours that might be associated with femininity. Numerous studies of female-dominated occupations and workplaces, whether factory work or secretarial work, have shown how women workers often draw on alternative, feminized scripts and performances in their discursive construction of identity and in their definition of work roles and relations with co-workers in order to challenge masculinized norms and male domination, introducing questions about familial obligations, pregnancy and menstruation, for example, into a previously disembodied workplace discourse. In studies of British and Japanese factory workers, for example, Westwood (1984) and Kondo (1990), respectively, have demonstrated how women in their workplaces assert their identities as mothers and as carers for others, paradoxically challenging masculinized dominance of factory spaces but, at the same time, reinforcing their own inferiority as workers by emphasizing their femininity and so reaffirming their status as 'other', different from the masculine norm. In a study of data entry workers in Barbados, however, Freeman (2000) has shown how female workers, by dressing well and emphasizing their femininity, are able to construct a discourse of work as significant and status enhancing, closer to the world of the office worker than routine semi-manual work, and so challenging their designation as unskilled.

Rosemary Pringle's (1988) analysis of secretarial work was one of the earliest and best of the studies that focused directly on questions of sexuality and desire in a service sector workplace, showing how sexuality, pleasure and desire are important in establishing and reinforcing workplace hierarchies. She defined sexuality as a set of meanings related to representation, identity and desire rather than explicit behaviours associated with sexual activities. Through interviews in offices, she showed how the relations between secretaries and their bosses in large part depended on gendered/sexualized interactions. Female secretaries, for example, drew on a range of gendered and/or sexualized discourses, including the office 'mistress' and the office 'wife', to construct particular power relations between them and their male bosses. While flirting and having fun was a common script for male boss-female secretary interaction, as well as the extension of office duties into 'homemaker' tasks (making coffee, arranging food at meetings, buying flowers for the office or gifts for the boss's household, for example), between women bosses and women secretaries the social relations were both more straightforward and less deferential and typically confined to more strictly defined office-based and work-related activities.

As Pringle argued, employees are not merely passive objects in these relationships. Instead, they are active agents who both construct and resist these different positions. Identities are not fixed as people enter the workplace, but are open, negotiable, shifting and ambiguous. Gendered sexualized identities are thus constructed and challenged through workplace practices in official and unofficial arenas and practices that are saturated by notions about gender and sexuality in ways that position women, in particular, as sexualized bodies. Thus:

> Sexual skills are acquired and incorporated into the organizational role. The organization acquires command over the sexuality of its employees, within certain limits. Women with jobs that require, implicitly or explicitly, an attractive appearance – hostesses, saleswomen, receptionists, secretaries – are duty bound to be agreeable or seductive, and must be or pretend to be 'sexy' in their dealings with the public. (Gherardi 1995: 43)

It is clear then that workplaces and organizations are not only (if at all) rational bureaucracies, but also locations and sites for the construction of identity, in which men and women 'do gender' (West and Zimmerman 1987) and construct and enact versions of sexuality through everyday interactions. Workplaces are dynamic and changing and are themselves embedded within wider social structures and attitudes and assumptions about gender and sexuality. Socio-spatial relations operating across different scales and times interact in the construction of workers' identities. Economic migrant workers, for example, may have difficulty in reading the scripts that structure sexualized performances in workplaces.

Theorizing Embodied Identities, Exploring Complexity

A growing number of studies have drawn on Pringle's path-breaking work, exploring the consequences of theorizing gender and sexuality as fluid and mutable in analyses of deferential performances in a range of occupations.

Expanding the definition of sexuality from sexual acts per se to include representations, everyday interactions and social regulations, as well as ideas of fantasy and desire, has opened up new areas of research about the economy. These studies have demonstrated how conventional attributes of hegemonic gender identity and a dominant version of heterosexuality are performed and confirmed in daily and institutional practices in workplaces in ways that benefit (certain) men. In the main, these studies draw on two connected sets of theoretical arguments. The first is Foucault's (1978) insistence that the body is an inscripted surface, in which self-discipline and normalization based on multiple discourses (temporally and culturally specific sets of ideas, images, institutional structures, practices and regulations) are crucial in the production and maintenance of an approved body. Thus through the operation of a wide range of forms of social regulation including self-discipline, 'conforming' or 'docile' bodies are produced and reproduced though everyday social practices. These bodies tend to conform to conventional social notions of acceptable versions of masculinity and femininity, but are open to redefinition through resistance to the norms.

The second set of work is that dominated by the arguments of the feminist theorist Judith Butler (1990, 1993, 1997), who draws on psychoanalytic, feminist and poststructuralist theories in her work on identity. She insists on the provisional status of identity which is performative, constructed within language and discourse. Gender is thus a 'regulatory fiction' constructed within discourses that normalize heterosexuality. Her notion of performativity is not one of an individualized and voluntary performance, but rather the creation of identities that are constructed through pre-existing discursive structures (Salih 2002). Through these structures, gendered identities are continually produced and reproduced, made material through schemes of surveillance, discipline and self-regulation. Thus, gendered identities are never singular nor fixed but fluid and complex, often contradictory, reiterative and performative, and so made and remade in social relations in the workplace. Masculinity and femininity are mutually but also multiply constituted, variable and relationally constructed, rather than being a categorically separate and unvarying binary division. Although gendered identities are context dependent and temporally specific they are nevertheless, as Butler insists, always inextricably embedded and produced within dominant representations of heterosexuality in western societies.

Gender, Butler argues, is routinely produced through 'a heterosexual matrix', a term that she uses 'to designate that grid of cultural intelligibility through which bodies, genders, and desires are naturalized' (Butler 1990: 151). In this grid, a hegemonic version of sex is established and through this a notion of gender that is 'oppositionally and hierarchically defined through the compulsory practice of heterosexuality' (p. 151). Normative heterosexual identities are maintained through the policing of hegemonic performances and the shaming of 'abnormal' performances, through a process of othering. Thus the very acceptance of 'intelligible' or normative gender identities depends on the contrasting presence of an abnormal or 'unintelligible' gender – the terms are Butler's. Consequently, heterosexuality and gender are inherently unstable as each relies on the contrasting presence of the Other. However, heteronormativity is dominant, set within a complex matrix of power relations that together constitute the hegemonic regulatory regime within a particular society.

As feminist analysts of the labour market have shown, this heterosexual matrix regulates social relations in the workplace, both through the social construction of certain jobs as appropriate for men or for women, and through the regulation of organizations and everyday social practices on the basis of heteronormative principles (Pringle 1988, 1998; Adkins 1995; Adkins and Lury 1996, 1999; McDowell 1997, 2003; Halford and Leonard 2006). In this regulatory system the embodied performance of an acceptable workplace identity is central. And although this matrix is malleable and open to rearticulation, sets of accepted constraints – both subjective and socially constructed discursive practices – limit the extent and possibilities of transgression. Transgressive performances are possible, however, challenging the association of masculinity or femininity with particular jobs and occupations – men doing 'female' jobs, for example, such as nursing or domestic service, that demand the sort of deferential or docile performance typically associated with femininity (Lupton 2000) or women undertaking masculine work such as heavy manual labour or rational calculating tasks. In chapter 7 the position of men in nursing and care work is explored. Adopting Butler's notion of the heterosexual matrix in analyses of the production of 'appropriate' working bodies thus allows interesting questions about employment practices to be addressed.

What these theoretical arguments allow is the linking of sexuality and gender to Wolkowitz's (2002, 2006) arguments about the body. In embodied, sexualized performances the attributes of a desirable and desiring body play a part. Attributes such as weight, complexion, hair, accent, clothes and gestures all become part of the interactions between providers and consumers. An employee in interactive occupations is usually trying to persuade the purchaser to buy something, be it a burger (Crang 1994), a hotel room (Adkins 1995; McDowell et al. 2007), financial advice (Halford et al. 1997; McDowell, Batnizky and Dyer 1997), a toy for a child (Williams 2006) or a ride at Disneyland (Van Maanen 1991). Some organizations make the requirement of an idealized, typically white, clean, slim and young, and often sexualized, body an explicit part of the recruitment process. When the Euro Disney Corporation was first recruiting in Paris in the 1990s, for example, a series of rules were laid down about appearance at work – no facial hair, no single earrings for women and none at all for men, women had to wear black tights and trousers could only be worn by women if they negotiated permission. Similar rules are common in the airline industry where the sexualized bodily performance of both men but especially women flight attendants is notorious, perhaps culminating in the sort of campaigns airlines were running in the 1980s and 1990s (Cathay Air ran advertisements in the 1980s featuring an attractive young woman with the tag line of 'I'm Cindi, fly me'). This emphasis on looks and age, on weight and appearance in the airline industry has been subject to legal challenges, however, and the age limit for working as part of the cabin crew has been raised.

In this industry, however, as in many forms of routine service work, interactions with clients are highly scripted, regulated and monitored and often depend on the manipulation of the employees' sexualized emotions, including the manipulation of sexual desire (Leidner 1993). In her now-classic study of a range of jobs including airline stewarding, Hochschild (1983) described these types of work as dependent on a 'managed heart' in which an affective relationship is established with passengers through the manipulation of emotions. Thus cabin crew produce a performance that combines deferential service with an authoritative knowledge of security and safety issues. In this performance, both hetero- and homosexual desire is a significant part of client-employee interactions and of interactions between workers. Du Gay (1996) has documented a different type of scripted exchange in retail outlets aimed at the teenage mass fashion market. In this case, a scripted exchange based on an ideal of youthful equality, rather than the heterosexist interactions Hochschild noted in the airline industry, is common. Further, the conventional distinction between the workers and clients is blurred in interactions that depend increasingly on the similarity of the sales staff and the customers and their participation in a sociable, yet scripted, ritual based on false notions of equality and familiarity. In these exchanges a groomed, trimmed, tamed and toned, sexually desirable body, preferably well dressed in the firm's products, and the capacity for continual self-discipline and improvement are a significant aspect of the employment relationship. Casual flirting is also a recognized part of the script, as it helps to sell clothes.

Grooming the body and dressing the part have long been acknowledged as central to the social construction of femininity. The hard work that is involved in producing an acceptable version of heteronormative femininity is nicely captured in this description of the female body by Andrea Dworkin (1974: 113–14):

Standards of beauty describe in precise terms the relationship that an individual will have with her own body. They prescribe her motility, spontaneity, posture, gait, the uses to which she can put her body....
  In our culture, not one part of a woman's body is left untouched, unaltered. No feature or extremity is spared the art, or pain, of improvement.... From head to toe, every feature of a woman's face, every section of her body is

subject to modification, alteration. This alteration is an ongoing, repetitive process. It is vital to the economy, the major substance of male-female differentiation, the most immediate physical and psychological reality of being a woman. From the age of 11 or 12 until she dies, a woman will spend a large part of her time, money and energy on binding, plucking, painting and deodorizing herself.

The constant attention to weight, odour, looks and appearance is not only the focus of individuals' daily routines but, as Dworkin notes, the basis of multi-million pound industries. And in the three decades since Dworkin described the construction of a woman's body, the body has become even more central to both individual identity and to the health of the economy. Bodily standards increasingly apply to men too, as they become the subjects and objects of advertising and marketing campaigns. For both men and women in the new millennium, the body is not only the subject of anxiety but also the last frontier of control as the lived experience of embodiment is increasingly an area of choice and modification through diet, exercise and surgery (Turner 2008).

The idea that body work *on oneself* is a central part of the new service economy is not a new idea either. Erving Goffman in his book *The Presentation of Self in Everyday Life*, first published in 1959, used metaphors from the stage in arguing that identity work involved performances, role playing, scripts and audiences in which individuals were seen as both the product and the producers of social meaning. He talked about the body as a peg on which something of 'collaborative manufacture will be hung from time to time' (Goffman 1959: 245), although in Foucauldian analysis the body itself is constructed through inscription: it too is part of the collaborative manufacture, rather than an unchangeable physical object as Goffman implied. The social construction of bodies is thus a social and interactive process in which individuals have agency (Featherstone, Hepworth and Turner 1991; Bordo 1993; Shilling 1993; Tseelon 1995). Although people are not necessarily the dupes of consumer capitalism – tricked into purchases and interventions against their will – they are nevertheless constrained by appropriate versions of embodied identities. And as more and more people engage in forms of work in which appearances matter, the body increasingly becomes the subject of self-improvement.

Writing more than thirty years after Goffman, British sociologist Anthony Giddens (1991) argued that bodily appearance had an even more central relevance than in previous generations. He acknowledged that dress and adornment have always been a signalling device of gender, class position and occupational status, but suggested that in late modernity – the period from the 1970s onwards – the very design of the body itself becomes an ideal to work on. No longer just a 'peg' as Goffman assumed, material bodies might themselves be reformed and reconstructed to achieve the desired – above all, youthful – appearance. Thus, he argued, 'bodily regimes are the prime means whereby the institutional reflexivity of modern social life is focused on the cultivation – almost, one might say, the creation – of the body' (p. 100) and so 'we become responsible for the design of our own bodies' (p. 102). Men, as well as women, are the subjects and objects of this redesign as the self becomes a project to be consciously and continuously 'worked at' (Gill, Henwood and Mclean 2005). Men's bodies are on display alongside women's on billboard, in films, in both popular and specialist magazines and as fashion icons. The image of a hairless, toned David Beckham (in an Armani advertisement in 2008), in tight white pants and splayed legs, revealing his splendid equipment as the object of both the hetero- and homo-erotic gaze, is a classic example of a male body coded in a way that gives permission for it to be looked at and desired by men as well as by women (Simpson 1994).

Like Turner, Giddens (1991) argues that this emphasis on the desired and desiring body leads to anxiety, insecurity and self-criticism in a society characterized by greater risks (this is a reference to Ulrich Beck's (1992) thesis about late modern society as a risk society) as the conventional structures of family, class and gender become more fluid and subject to renegotiation by individual action. Thus, in parallel with Susan Bordo's (1993) work on eating disorders and the female body, he argues that the rise of eating disorders among young men as well as young women, as well as other forms of body hatred and body dysmorphic disorders and self-harming, is in part a consequence of the focus on embodiment in contemporary society:

> Anorexia and its apparent opposite compulsive over-eating, could be understood as casualties of the need – and responsibility – of the individual to create and maintain a distinctive self-identity. They are extreme versions of the control of bodily regimes which has now become generic to the circumstances of everyday life. (Giddens 1991: 105)

Naming workers

In interactions with clients in the work spaces of service economies, the expectations that the customers hold about the ways in which a service should be provided and who is a suitable worker enter into decisions not only about the performance of the task but also who should be appointed to do it. In service labour markets there is a triadic or three-fold relationship involved in workplace practices: the key actors are managers/employers, workers and customers. Williams (2006), in her case study of the ways in which customer expectations affected recruitment policies in retail outlets in two localities in a large US southern city, showed how class and ethnic differentiation between the areas was reflected in employment practices as the outlets recruited staff who mirrored the different expectations of the residents in each area. In her work, Williams (2006) drew on the concept of interpellation. Interpellation (call and response) is an Althusserian concept, applied to labour market analysis by Michael Burawoy in his book *Manufacturing Consent* (1979) to capture the ways in which employers/managers construct idealized or stereotypical notions of idealized workers. This naming of others in the workplace is in turn internalized by workers themselves so that they come to conform to or recognize themselves in the managerial naming. Thus subjects are constituted in and take meaning from social relations in the workplace. Identity is not an inherent attribute of the individual, but a social construction. Workers who come to embody managerial assumptions/stereotypes – about docile femininity, for example, or embodied masculine strength – are in part conjured up by managerial fantasies.

In recent work, the concept of interpellation has been extended in its confrontation with the feminist studies I outlined above, and their recognition that identities are more fluid and malleable as well as multiple than earlier labour market theorists suggested. Salzinger (2003) and Williams (2006), for example, drawing in particular on the work of the feminist film critic Teresa de Lauretis (1987), have insisted, unlike Burawoy, that workplace identities, which he saw as constituted only in and by class divisions, are not singular but multiple, and indeed may be contradictory, the site of resistance as well as conformity to managerial namings. Wright (1997, 1999, 2006), for example, in her analyses of women's manufacturing work in Mexican macquiladoras (factories on the US/Mexico border), showed, like Salzinger (2003), that Mexican women in different circumstances are able to challenge their construction as 'Woman', as stereotypical docile female subjects, through a range of workplace strategies. Thus Wright suggests that interpellation is a *contested* process, 'paralleled by strategies of resistance, as workers challenge the dissonances between their own desires and self-identities and managerial/client expectations' (Wright 2006: 56). It is important, however, also to recognize the multiple discourses of managers and employers and not to see these as singular or unchanging stereotypes. They too are located within organizational structures that produce and reproduce certain versions of managerial discourses that emphasize particular attributes of desirable future employees. Thus workers are 'formed in dialogue with other workers ... through comparison, contrast and opposition to multiple imaginaries' (Salzinger 2003: 20), although as Williams (2006: 55) noted, 'workers typically consent and embrace the stereotypes (*employed by management*), since their opportunities depend on their conformity to these managerial imaginings'. As Bourdieu (1999) argued about a form of aggressive masculinity that characterizes young working-class men's sense of their identity, it is often less painful 'to make a virtue out of necessity' (p. 433) than to challenge stereotypes, to resist categorization or change behaviours.

While Salzinger and Wright's case studies were of manufacturing workplaces, Williams' study was of service employment, where customers, as well as managers and co-workers, also construct a series of imaginaries in anticipating service interactions. In service jobs and occupations interpellation takes what Williams termed a *dual form*, as workers not only have to conform to managerial imaginations of an idealized embodiment of service but also to the expectations of customers, from airline customers who expect service with a smile to accompany a speedy check-in process, to the guests of the hotel who want efficient but authoritative service in the restaurant and the invisible servicing of their rooms when they are unoccupied (Waldinger 1992; Guerrier and Adib 2000). Front-stage service workers (Goffman 1959) are thus the visible objects of the multiple desires and fantasies of clients who not only purchase a service but also a set of expectations – whether about luxurious pampering and 'time-out' or an efficient business service. Indeed, as Gabriel (2004), Ritzer (1999) and others (e.g. Bryman 1999, 2004) have argued, in consumer-based industries, 'enchanting' the clients has become a key part of service provision, and service workplaces become 'more oriented towards the "fantasizing consumer" than the "toiling worker"' (Hughes 2005: 609).

In high-tech and high-status occupations the sexed body and sexual desire are paradoxically both present and absent. Although an aestheticized, sexually attractive and sexually conformist (heterosexual) embodied and interactive performance by individual employees is highly valorized (Bauman 1998), the status of these types of occupations depends on their construction as cerebral and disembodied: attributes which are, of course, traditionally associated with a particular version of hegemonic masculinity (Connell 1995, 2000). In the idealized version of bureaucratic or scientific work, the worker is a rational, calculating instrument, free from the messy emotional demands of everyday life and, rather like a medieval monk or old-fashioned Oxbridge academic (Massey 1995), freed from the need even to provide his (and the model is a masculine one) own meals. In this version of the working world, relationships are based on reason and not emotions which are seen as inappropriately intrusive in the workplace. However, in the world of 'soft' capitalism identified by Thrift (2005), new sets of relationships within organizations based on ideas of coping with complexity and uncertainty and the need to produce learning environments in knowledge-based organizations have become significant, at least rhetorically. In devising structures to deal with complexity and uncertainty, the need to 'engage hearts and minds' (Thrift 2005: 32) became significant, almost as if it was suddenly recognized that 'organizations were made up of people after all', not just 'heads' and 'role occupations' (Handy 1989: 71).

Ideas about pleasure, emotions and embodied or tacit knowledge began to pervade the management literature as organizations searched for ways to deal with uncertainty and diversity and to build trust in different circumstances. Management turned to a range of different sources to help high-level workers 'get in touch with themselves', including, in the 1990s, New Age ideas. A range of organizations from the Bank of England to the large insurance company Legal & General sent top executives on courses to learn the Whirling Dervish dance in search of inner peace (Thrift 2005: 42)! The management literature and practices, however, developed in almost complete isolation from feminist analyses of the significance of embodied emotions and sexuality at work, drawing on Polanyi's (1967) ideas from gestalt psychology (his most famous saying was 'we know more than we can tell') rather than the work of theorists who were more influential in feminism and in the theoretical social sciences, such as Bourdieu (discussed below). However, it is clear that traditional ideas about disembodied cerebral practices in the workplace were being challenged in practical ways in many organizations, as embodied emotions were admitted into the workplace, even into the board room.

Unlike organizational theorists, feminist philosophers looked elsewhere for ideas about the significance of the lived body. In a review of the work on embodiment, US philosopher Iris Marion Young (2005) argued that an analysis of social structures must be added to understandings of how normative heterosexuality constructs/positions different bodies. As she argues, 'social structures position individuals in relations of labour and production, power and subordination, desire and sexuality, prestige and status. The way a person is positioned in structures is as much a function of how other people treat him or her within various institutional settings as of the attitude a person takes to himself or herself' (2005: 20–1). This chimes with the Althusserian notion of interpellation and is a further reminder that in service economies, the attitudes (or assumed attitudes) of customers, as well as employers, are significant in explaining who gets what sorts of work. These attitudes, however, are constructed within social structures. The categorical inequalities of class, gender and ethnic origin still matter. But Young also reminds us that individuals may occupy multiple positions in structures and so different attributes and positions are salient in different arenas of life. In her book *Justice and the Politics of Difference* (1990), which is a theoretical critique of notions of distributive justice, she explores the connections between the division of labour and embodiment, arguing that contemporary structures of inequality are composed of what she terms 'five faces of oppression', involving 'social structures and relations beyond distribution' which create hierarchical differences between groups of people (p. 9). The five faces of oppression she identifies are exploitation, marginalization, powerlessness, violence and cultural imperialism. Cultural imperialism is the face that is most relevant here and perhaps the least self-explanatory. It is defined by Young as the 'universalization of the dominant group's experience and culture, and its establishment as the norm' (p. 59), which works to stereotype the views and experiences of people outside this dominant norm and construct them as 'Other'.

One of the key aspects of cultural imperialism involves the body or bodily image and its presentation. As Young argues, ugly, fat, non-white, elderly bodies are inadmissible in societies that valorize an idealized white, slim, young, unwrinkled, typically heterosexualized body, and so such bodies are out of place in the interactive sales/advice-giving industries and occupations that increasingly dominate in advanced industrial societies. In these consumer societies, 'dynamics of desire and the pulses of attraction and aversion' (Young 1990: 60) influence the scope and content of interactions between workers, their peers, superiors, clients and customers, a reaffirmation of the earlier argument about the significance of consumers in service economies in expecting to be served by desirable bodies. The ability to achieve the most desired image is constrained not only by income and resources but also by ageing and bodily decline, as well as the extent to which diverse bodies can be manipulated to achieve what is most highly valorized. And the bodily ideals presented in television and film and representations in magazines and in advice columns are themselves typically unattainable: not achievable identities but rather fictions (Frost 2005) or the simulacrum identified by Burawoy (1979).

The centrality of desire to the economy as whole, as well as to workplace performances, has been addressed by Zygmunt Bauman (1998) in his assessment of the consequences of the rise of consumer societies. He argues that the transition of advanced industrial economies from mass societies, mass producing a limited range of consumer goods for relatively undifferentiated markets, to economies based on the development of a highly differentiated range of products for niche markets has, in large part, been based on the successful manipulation of desire, associated with the rise of advertising. He argues that as consumer societies are volatile and temporary, objects of desire not only must be instantly available but must also bring instant satisfaction which just as quickly wanes. 'Consumers must be constantly exposed to new temptations in order to be kept in a state of constantly seething, never wilting excitation and, indeed, in a state of suspicion and disaffection' (p. 26). In an even more explicit parallel with sexual desire he argues:

> It is often said that the consumer market seduces its customers. But in order to do so it needs customers who are ready and keen to be seduced. In a properly working consumer society consumers seek actively to be seduced. They live from attraction to attraction, from temptation to temptation. (p. 26)

The rapid gratification of consumer demands is thus a driver of economic expansion, as well as a central element in the construction of self-identity.

The purchase of 'lifestyle' choices becomes a way of marking out both distinctiveness and distinction (Bourdieu 1984), as affluent consumers are increasingly able to buy exactly what they need to mark themselves out from others in a 'society organized around desire and choice' (Bauman 1998: 29). Almost everything is for sale. Customers are able to buy a 'facelift, Armani suit, liposuction, phalloplasty, Porsche, blow job, a whipping or bondage session ... to enhance their self-esteem in the most appropriate way' (Hawkes 1996: 117). And so in this type of consumer-based society the role of the providers and sellers of goods and high-quality services in niche markets as agents of seduction assumes a highly significant role. In growing numbers of service occupations, the interaction between clients and providers is an exchange based on the manipulation of emotions and the satisfaction of desire, and in these transactions the bodily performance of the server is a key part of the exchange: whether literally or as a symbolic representation of the service. From sex work, through elder care to university teaching, an embodied performance is a central part of the exchange.

Clearly, the recognition of the ways in which gendered identities are a key part of workplace social relations has been central to the work that I have summarized so far. However, bodies are also differentiated by age, by sexual preference and by different levels of physical ability, as well as by social class. As Robyn Dowling (1999) noted in an assessment of geographical work on the body, until that date there had been an almost complete neglect of the fact that bodies also have class positions written on to them as well as gendered attributes, although she ignored Burawoy's use of the concept of interpellation to explore class-based practices. In the last part of this chapter, I consider other dimensions of embodiment and the ways in which they interact in the labour market to produce hierarchies of desirable and appropriate bodies for different types of work.

Class Practices and Ethnic Penalties: Recognizing Complex Intersectionality

So far in this discussion about embodied labour and the construction of identities in the workplace, the focus has been on gender divisions and gendered social relations, in part because the associations between femininity and emotions, gender and embodiment are so obvious and significant in understanding new divisions of labour. But bodies bear the traces not only of gender relations but are also marked by class and ethnicity and the sets of assumptions that accompany these differences. In feminist theory and in the work on embodiment more generally, it is now widely recognized that identities are complex, multiple and fluid, continuously (re)produced and performed in different arenas of everyday life. Responding to criticisms that gender is not a single category, untouched by class position or skin colour, feminist theorists now insist that subject formations and social relations are constructed through *intersectionality* – a set of relationships among the multiple dimensions of being. Thus, identities are theorized as complex and diverse and, in a formulation that appeals to geographers, as historically variable and spatially contingent – in other words, time and place matter (McCall 2005). The dimensions of difference typically referred to in discussions of complex intersectionality include race, gender and ethnicity; class, sexuality and age are sometimes addressed too and there are now numerous, often empirically sophisticated, studies that focus on connections between these dimensions in explorations of labour market change. Typically, these analyses draw on qualitative research methods, unpicking connections through careful and detailed interviews or ethnographic research with small groups of individuals or communities, showing how men and women of colour, workers with alternative sexualities and older workers find themselves disadvantaged in particular labour markets, disqualified from some positions and crowded into others, typically at the more precarious end of the labour market where rewards for work are lower than average (see, for example, Benaria and Roldan 1987; Ong 1987; Hanson and Pratt 1995; Kim 1997; McDowell 1997; Chatterjee 2001; Salzinger 2003; Chari 2004; Chari and Gidwani 2005; Wright 2006; for an excellent review, see Mills 2003). To end this chapter, I want first to discuss the connections between class, embodiment and employment and then turn to questions about ethnicity and skin colour in the labour market.

Class practices

The production of class as a location in the sphere of production typically has been theorized as the outcome of economic change rather than as part of its explanation. Class, it is argued, is an objective location. It consists of structurally defined categories, which in Marxist theory are constituted by a relationship to the means of production. Capitalist societies in their simplest form consist of two classes with opposed interests: the owners of the means of production – the bourgeoisie – who profit from the labour of the working class who must sell their labour power in the market to exist. As societies become more complex, new divisions are created between, for example, owners and mangers who control the labour force on their behalf, but nevertheless class is seen an a consequence of rather than an input to economic relations. Interestingly, Marx himself did not ignore the body. Rather, he argued that there is a mutual and constitutive relationship between the body and work: the body is both the source of labour and its output, altered in different ways by the labour process. In a telling phrase Marx defined the outputs of employment – goods in particular – as the 'memorialization' of embodied work. He also argued that the tools used by labourers might be seen as an extension of the body. These arguments are perhaps easier to relate to the types of heavy manufacturing employment that were growing as Marx wrote in the nineteenth century, although service sector work in finance and business, for example, was also growing rapidly at the same time. In service work, where the output is often ephemeral, weightless or used up in the exchange, the labouring bodies are not memorialized in concrete form but rather in reputation, enjoyment or the prospect of future contacts. And as I argued earlier, Marx himself ignored service employment.

But class is more than a structural location, as the recent turn in class analysis to analysing the sets of cultural practices that construct and reinforce class differences demonstrates (Cannadine 2000; Savage 2000; Devine 2004; Skeggs 2004a; Devine et al. 2005; Sayer 2005). Here the work of Pierre Bourdieu has been helpful in understanding how ways of living, consumption practices and embodiment are all part of the production and maintenance of social divisions, positioning some people as superior to others in terms of their attitudes, beliefs and ways of living as well as in their occupational status and standard of living. In *Distinction* (1984), Bourdieu explored the ways in which the French middle class distinguished itself from the working class through the ownership of sets of belongings and lifestyles that defined their owners' moral superiority. Bourdieu used the term *symbolic violence* to capture the ways in which certain groups were constructed as morally inferior. Symbolic violence is different from physical violence: it is a more insidious form of power, constructed in both laws and practices of social institutions and everyday life. As Bourdieu (2001: 38–9) notes, 'it is a form of power that is inculcated on bodies, operating at "the deepest level of the body", and its efficacy is derived from the fact that it continues to exist long after external forms of violence are removed.' He argued, for example, that women's increased access to occupations, formal education, the political sphere and the right to vote cannot completely 'undo' the effects of the internal barriers that are imposed on women by acts of symbolic violence. In his view, a process of 'self-exclusion' takes over from 'external exclusion', thereby perpetuating male domination over women (Bourdieu 2001: 39). These arguments parallel Foucault's notion of disciplinary power and Butler's heterosexual matrix.

Bourdieu's (1984, 1990, 2001) work is insightful about the connections between embodiment, class position and class practices. He demonstrated the ways in which the body is marked by class signifiers and practices – made visible in the way in which people stand, in their gestures and habits, as well as through the ways they speak and dress. Thus the disposition of the body, he argued, is a social not a natural phenomenon: subjectivity is constructed through a person's location in a social field or set of social relationships – a feature that Bourdieu captured in the term *habitus*. Habitus is thus the set of structured and structuring relationships that frames bodily conduct. This conduct itself Bourdieu termed *hexis*. As Bourdieu (1984: 466) recognized, 'social distinctions and practices are embedded in the most automatic gestures or the apparently most insignificant techniques of the body – ways of walking or blowing one's nose, ways of eating and talking'. In social interactions, one of the hallmarks of practice is the way in which the appearance of the body is significant. Assumptions and connections are made that occur without the explicit intervention of a discursively based consciousness or reflection. As Bourdieu has argued, 'there is a logic that unfolds directly in bodily gymnastics' (1990: 130). This claim parallels more recent discussions in other disciplines (Ahmed 2004), including geography (Davidson, Smith and Bondi 2005; Tolia-Kelly 2006; McCormack 2007), about the significance of 'affect' and emotions in social exchanges. Almost instantaneous, typically unreflexive reactions based on visceral emotions label and categorize actors in particular social circumstances. These emotions, as Hans Gerth and C. Wright Mills (1964) argued decades ago, are based on an inextricably intertwined combination of gesture, physiological processes and conscious experience. And so, social actions are always embodied just as embodiment is constructed socially. 'The social is incorporated into the body' (Skeggs 2004: 5) and vice versa.

Furthermore, embodiment is fluid and changeable. As individuals move between different social arenas, their habitus may change as they acquire new sets of skills and social dispositions over their lifetime, building on what Bourdieu terms their social, economic and cultural capital. Social capital is generated through family position and its relationship to wider society and is constituted largely though social networks, bringing useful contacts for some and excluding others from the networks that confer social advantage and privilege. Economic capital is wealth, either inherited or earned as income through interactions between individuals and economic structures. In an interesting comparison of young women in the US, Fernandez Kelly (1994) showed how the differential possession of social and economic capital of girls of colour growing up in the ghetto differentiates and disadvantages them from the moment of their birth. Cultural or symbolic capital is that set of social relations manifested in status and prestige, and interpersonal qualities such as charisma (Bourdieu 1984). Clearly, these capitals are interconnected and constructed over the lifetime through participation in different social arenas. These capitals all have an economic value if they are in short supply and so sought after. Thus attributes of social capital ranging from an elite education to modes of thinking or qualities of style are differentially evaluated and have different values in the labour market, constructing and maintaining patterns of social and economic exclusion (Brown 1995), as I illustrate in later chapters.

Reflexivity and individualization

In an interesting series of arguments, Lisa Adkins (2000, 2003, 2005), a sociologist interested in gender and work, has brought together a critique of the arguments about knowledge economies as increasingly individualized with Bourdieu's arguments about habitus and field to suggest that one of the consequences of service sector growth has been a reconfiguration of gender relations rather than their decreasing significance as Ulrich Beck, among other theorists, has suggested. Beck, a German sociologist, is one of the most influential commentators on recent social changes in industrial societies. He argued that western societies have become risk societies (Beck 1992) in which the traditional mechanisms of class solidarity and social movements to ensure security have been destabilized both by greater risks – of famine, disease, war, nuclear threat – and by a shift in economic and social policy towards a neoliberal version of individualization or individual responsibility. Thus, Beck argues, the traditional constraints of the former industrial society are weakened. The new post-industrial world is one which is distinguished by self-reflexivity – 'the ability to think and reflect on the social conditions of … existence and to change them accordingly' (Beck 1994: 174). In work with Giddens and Lash, Beck (1994) has dubbed the current era one of 'reflexive modernization' in which new opportunities for social (and physical) mobility have emerged. People – social agents – have greater freedom to experiment, to move and learn, and to take different jobs over their lifetime than in industrial economies. The most successful are people who are able to build 'portfolio' careers, able to construct and sell their individual experiences in the knowledge-based economy in which performance, style and confidence are as important a part of working life as more traditional skills. Categorical inequalities (Tilly 1998) – the structural constraints of class and gender – are becoming less significant as notions of individual rights and responsibilities become more important. Thus Beck and Beck-Gernsheim (1996: 29) have suggested that 'people are being released from the constraints of gender … axes of difference, such as class, gender and sexuality (even life and death itself), are more a matter of individual decisions'. This claim seems astonishing on first reading, although it is evident that new technologies such as post-menopausal assisted conception and assisted suicide support it. I want to argue below, however, that class and gender and ethnicity – the traditional axes of difference – retain their salience in the contemporary workplace.

This optimistic theoretical scenario about the significance of reflexivity nicely parallels the focus on individual effort in the neoliberal economic and welfare policies that characterize the British state at present, as well as the business rhetoric about the need for flexible workers, able to reskill and retool as the economy demands. It is, however, important to note that Beck

(2000), like the theorists discussed in chapter 2, also recognizes growing polarization in the new economy. Indeed, in a somewhat apocalyptic scenario he identifies the potential 'Brazilianization' of western economies as the self-reflexive elite live lives increasingly separated from the urban dispossessed, trapped without work or in poor jobs that bring insufficient income for a decent standard of living. It has, however, been the optimistic side of Beck's arguments that has appealed to many theorists, seeing opportunities for the 'detraditionalization' of new knowledge economies.

Adkins (1995, 2000, 2003, 2005) disagrees with these optimistic assessments, arguing that these opportunities for remaking life remain strongly gendered as gender identities seem stubbornly resistant to reflexive rethinking. She suggests that a new gendered binary is being constructed in dominant theoretical explanations of the post-social era. Drawing on Lash's (1994) arguments about life chances in the post-social era – what Lash refers to as the structural conditions of reflexivity in which access to the mode of information replaces access to and place in the mode of production – Adkins argues that men are what Lash (1994: 133) termed the 'reflexivity gainers' and women the 'reflexivity losers'. Women are excluded from the higher ends of the high-tech information and knowledge economy, becoming part of a new under or lower class in the polarizing labour market. It is in this class position, Lash suggests, that the 'ascribed characteristics of "race, country of origin or gender"' (Lash 1994: 134) are of crucial significance. So while men become the active subjects in the new mobile post-social order, many (most?) women remain trapped in the social, repositioned as they enter the labour market in growing numbers and yet trapped within a location that mirrors the old class and gender hierarchies of industrial societies because of their exclusion from the cultural field that valorizes reflexivity and mobility. Thus in Lash's work, Adkins (2004: 147) argues, women are located 'as *overdetermined* by the social and men as freed from the constraints of the social, or at least from the constraints of socio-structural forms of determination' (original emphasis). Ironically, Adkins notes that Beck, in a belated gesture to feminist arguments that previously he had ignored, now insists that 'gender is part of an older modernity and moreover women find it difficult to remove themselves from these social traditions and become individualized subjects' (Beck 1992: 151). So women are now trapped in the social rather than the pre-social, as regulation theorists such as Aglietta and Lipietz argued (see my critique in McDowell 1991), forever doomed, it seems, to lag behind men's progressive march forward to individual freedom, and so unable to 'achieve the form of personhood required to participate in the new modernity' (Adkins 2004: 152). Women are still committed, presumably by their consciences, if no longer by oppressive gender relations requiring them to remain in the home, to a version of freely given care and love for their dependants, unable to become the mobile individualized subjects of a new post-social networked space. As Adkins (2003: 28) notes, 'theories of reflexive modernization run the risk of reinstating the disembodied and disembedded subject of masculinist thought', leaving women undertaking emotional labour in different spheres much as usual, despite their large-scale entry into the social relations of waged work. As the chapters in part two document, women are the central component of the workforce undertaking emotional labour and caring work, as well as performing the same sorts of functions for 'love' in the home.

While Adkins draws on, or critiques, the theories of Beck and Lash, as well as Bourdieu, other feminist theorists (Skeggs 1997, 2004a; McNay 2000; McRobbie 2004; Moi 2005) have usefully built on Bourdieu's work in a different way – to analyse the ways in which social class divisions *between* women are increasingly marked on the bodies of women, especially working-class women, through the mechanism of symbolic violence. Skeggs, for example, has documented the recent rise of class antagonism and a discourse of moral superiority in the UK that position working-class bodies as morally inferior, as too big, too loud, too present (Skeggs 2004a, 2004b). At the turn of the twenty-first century, rather than being seen as the repository of decency and industrial solidarity, the British working class is now discursively constructed as unmodern, anti-cosmopolitan, backward and worthless, not playing its part in the newly competitive and multicultural Britain: defined and denigrated by what Haylett (2001) has argued, drawing on Bourdieu, is a form of class racism. Further, as the economic dependence on interactive service employment deepens, bodily performance has become increasingly significant in the lived practices that constitute and reinforce class antagonisms (Adkins and Skeggs 2004). Working-class bodies are marked as increasingly unacceptable in the tanned, toned world of the service economy and in commodified forms of consumption and entertainment (Young 1990; Wolkowitz 2002). McRobbie (2004), for example, has analysed the rise of reality TV 'make-over' programmes in which (usually) unattractive, fat, working-class or otherwise abject women are shown ways of transforming their embodied selves by so-called experts in the hope that the acquisition of new forms of social and cultural capital will improve their status and their life-chances. As McRobbie points out, these programmes typically involve a range of interactions that can be defined as symbolic violence, from a strict ticking off for poor posture, inadequate diet or bad hair, to outright sneers and on occasion outright humiliation of the participants. Thus McRobbie suggests that these programmes 'actively generate and legitimate forms of class antagonism particularly between women in ways that would have been unacceptable until recently' (p. 100).

This growing disdain among the more affluent for the less fortunate, and indeed for what are popularly seen as working-class attitudes more generally, is a more widespread phenomenon than in reality TV. Other examples include the adoption by the media of mocking discourses, labelling working-class youths as 'chavs' – a lumpish youthful proletariat distinguishable by their clothes and jewellery. Indeed, young people wearing sports-tops with hoods – the eponymous hoodies – were banned from a shopping centre in southern England in 2006, solely on the grounds that their dress reflects expected anti-social behaviour. The so-called 'respect agenda' and the continual use of the term 'yob' in discussion about the lack of respect in contemporary Britain is another part of the demonization of working-class youth (McDowell 2006b). Ferdinand Mount (2004), a right-wing 'one-nation' conservative, has argued that a new class divide is emerging in contemporary Britain based in cultural attitudes and divisions that are far wider than in earlier generations, as new forms of consumption and new methods of defending space emerge that increasingly act as markers of distinction and/or of disdain. Interestingly, Mount identifies the growth of a new form of cultural condescension towards the masses among the ruling class, in some ways paralleling Zygmunt Bauman's (1998) arguments about the growing significance of class divisions based on consumption practices rather than labour market divisions.

Recent evidence of this class condescension, or perhaps what might more accurately be termed a discourse of moral disapproval, has been identified by class theorists as well as in popular texts and the media. Sayer (2005) and Skeggs (2004a, 2004b), for example, both argue persuasively for the inclusion of a cultural and moral dimension in the construction and analysis of contemporary class divisions. Skeggs (2004b) has suggested that the significance of visceral emotions has been underestimated in explanations of class divisions. Middle-class disgust and resentment are, she argues, key factors in contemporary forms of class representations and class divisions. This disgust is, in large part, reflected in judgements made about the inappropriateness of the bodies of others, and so is of growing significance in the types of jobs that are currently expanding at the bottom end of the economy.

Racialized Others

I want to conclude this argument about the significance of embodiment with a discussion about skin colour, race and ethnicity and their connections to judgements about acceptable labour market performances, drawing parallels between racialization and the construction of gendered identities. There is a huge literature documenting the 'ethnic penalties' suffered in the labour market by people of colour (Heath and Cheung 2008). However, as most of this literature takes for granted rather than problematizes the associations between skin colour/nationality and evaluations of the worth or appropriateness of particular embodied performances in the workplace, I have ignored it here. Instead, I want to consider the ways in which ideas about subjectification developed by Foucault and Butler might be applicable to understanding the 'raced' as well as the gendered body at work. While the arguments that I outlined above about the construction of sexed identities as lying in the repeated regulation of the norms of sexual difference are now widely accepted, Butler and other theorists of performativity and subjectification have been largely silent about how to theorize the relationships between gender and other dimensions of social difference, while not denying the need to think about these connections. Butler (1990: 3), for example, recognizes that 'gender intersects with racial, class, ethnic, sexual and regional modalities of discursively constituted identities'. Indeed, she claims that 'as a result, it becomes impossible to separate out "gender" from the political and cultural intersections in which it is inevitably produced and maintained' (p. 3). However, her belief in the contingent foundations of all identity categories is less fully realized for categories other than gender. How does gender interact with ethnicity and nationality, with skin colour or social class, for example, in thinking about who gets what forms of work for what reasons in interactive service employment? Are other differences also performative, constructed through multiple systems of regulatory norms that are established in discourse? And how are they connected? There has been little agreement among feminist theorists, for example, about how to address these questions of complex identities, of the intersections of class and ethnicity with gender, other than a desire to move beyond earlier notions about the serial or additive construction of difference (Flax 1990; Alexander and Mohanty 1997; hooks 2000; Mohanty 2002). But the consideration of the position in the division of labour of people of colour and economic migrants (the two are not necessarily coincident) demands the theorization of ethnicized and racialized differences, as well as gendered identities and class positionality and practices.

In Britain and the USA it might be argued that the black/white binary division acts in a similar way to gender in making subjects and making workers. In these societies whiteness is the hegemonic norm against which the abnormal Other is defined. It is now widely accepted that race and ethnicity are also socially constructed categories. The purported attributes of inferiority that are mapped on to skin colour in these countries parallel the way in which gender is constructed as a binary distinction in which superiority is associated with masculinity and inferiority with femininity. A similar binary division is evident when white and black are compared and evaluated. Whiteness is constructed both as the unmarked norm and as conferring superiority on those with white skins (Frankenburg 1993, 1997; Dyer 1998; Bonnett 2000). Whiteness connotes goodness, 'all that is benign and non-threatening' (Dyer 1998: 6), in comparison to the darker skins of multiple Others, typically lumped together as Black. However, an interesting set of historical studies in the USA has documented the ways in which whiteness is also fluid and performative rather than a categorical distinction. Ignatiev (1996), for example, documented the way in which Irish migrants in the USA became white as Black workers moved from the South to the North; Roediger (1991, 1999) has documented a parallel process among European Jews and later migrants from Southern and Eastern Europe who also became white (enough). In comparison, more recent migrants to the USA from South and East Asia, for example, found that they were Black (Ong 2003). In the UK too, a black/white binary distinction was established as immigration began to increase, especially after the end of the Second World War (Winder 2004). The majority of in-migrants to the UK in the second half of the twentieth century were subjects from ex-colonies and on entry to the UK they found, to their surprise and/or shame, that they were uniformly regarded as Black (Fryer 1984). A stark binary division replaced the status hierarchies based on skin tone – termed pigmentocracies – found both in the Caribbean and in South Asia (Hall 1995): a distinction that is still significant. As Freeman (2000) found in her study of women's work in Barbados, a light(er) skin confers superiority in the labour market.

As Ong (2003) has shown in her analysis of the position of Koreans in the USA, hierarchical schemes of racial difference develop that include ideas about difference and the right to belong (or not) to the receiving nation. Racialized differences are also constructed through the attribution of a range of other despicable or inferior characteristics to visible Others. Discourses of racialized differences intersect, for example, with heteronormative constructions of gender as racial and ethnic markers are deeply gendered. In terms of suitability for different types of employment, examples include discourses about the different skills and talents that are apparently 'natural' attributes of some groups. The attributes are both racialized and gendered and so construct (in)appropriate bodies in different arenas. Ideas about different national work ethics, differently sexualized bodies, and about different roles in family and household, all affect options in the labour market (ideas of good wives, daughters, significance of family life, etc.) and coincide to confine migrants/people of colour to particular inferior spaces as both workers and (potential) citizens. Thus newcomers from different parts of the world and their descendents born in the USA and the UK are judged/placed within given schemas of racial difference, civilization and economic worth which substantially restrict their labour market opportunities. Stereotypical representations of Black men and women as erotic and carefree, of Black women as fecund and Black men as threatening or sexual predators, of Asian men as feminized, and Asian women as ultra-feminized and as caring and loving, are common in both North America and the UK (Webster 1998; Paul 1997; Pratt 2004; Kelly and Moya 2006): all of which map onto ideas about appropriateness for different types of work in the service economy.

For people of colour and migrant workers, bodily presentation also often confirms their social construction as 'less legitimate' (Bauder 2006). Accent, dress, self-presentation, behaviour, skin colour, hair, jewellery and height, especially in service-dominated economies, are used to mark migrants/minority populations as inferior subjects and workers, in jobs where the white heterosexual, slim body is the hegemonic ideal (Young 1990) and where 'human capital, self-discipline and consumer power are associated with whiteness' (Ong 1996: 739). Insights from post-colonial theorists add to explanations of how these discourses and markers of difference operate in western economies. Migrants from third world societies, for example, are constructed through the colonial gaze as Other, as backward and non-modern in comparison to the subject of western modernity (Bhabba 1994; McClintock 1995; Young 1995; Spivak 1999), and second-generation minorities continue to bear these markers on their bodies. Thus, the complex intersectionality of gender, class and ethnicity/skin colour produces subjects who are coded as inferior through the operation of numerous binary and categorical distinctions that are, through discursive practices, made complex, but also challenged, resisted, altered and transformed.

In post-millennium Britain, new questions about the location and ethnic identities of migrant workers are being raised as post-imperial subjects are being displaced/replaced as low-wage workers at the bottom end of the service sector by white-skinned migrants from Eastern Europe. The rapid rise in the number of migrants from the 'new' Europe on the accession of ten new states (Cyprus and Malta plus eight former communist states – Poland, Latvia, Estonia, Lithuania, Hungary, the Czech Republic, Slovenia and Slovakia) in 2004 and a further two (Bulgaria and Romania) in 2007 has added a considerable degree of complexity to the debates about the connections between nationality, skin colour, class and gender and the suitability of differently produced bodies for different categories of employment in the contemporary British labour market. New discourses of (lack of) worth based on national stereotypes, as well as the common European heritage and white skins of these new migrants, have disrupted the previously relatively stable connections between migration, ethnic origins and skin colour established since 1945 (Paul 1997; McDowell 2005, 2007).

It is clear that constructions of difference – whether based on class, gender, race, nationality, language or skin colour – are produced and maintained through discourses and practices that operate at and across different spatial scales and have come to have growing salience in a new service economy where personal interactions are a crucial element of labour market exchanges. The practices that construct and maintain difference include ideological assumptions, multiple regulatory systems, structures of power and domination, and spoken and enacted everyday practices in multiple sites, operating at both conscious and unconscious levels. While these practices are discriminatory and exclusionary they are also open to contestation and renegotiation. Thus a range of issues about the connections between embodied difference, labour market practices and challenges to contemporary structures of oppression and inequality arise that can be answered only by empirical analyses, drawing on multiple theoretical perspectives, from poststructural theories of identity formation to structural explanations of economic change. In the next sections of this book I turn to an analysis of some of the exciting empirical work on service sector occupations that has been undertaken in the last decade, illustrating the diverse ways in which working bodies are constructed as suitable for different types of work in the new service economy. Although the focus is on the UK in many of the examples, the scale and pattern of global migration mean that the workers themselves increasingly are part of the global division of labour, often born elsewhere and, increasingly, working for transnational firms, even when employed in the most local of jobs, perhaps providing cleaning services for the local hospital or caring for elderly Britons in their own homes. I have organized the chapters, rather loosely I admit, by scale, as well as by a number of other factors including a notion of confinement in terms of where the work takes place and of (un)willingness to perform it, as well as by skills and social status.

In part two there are three chapters that focus initially on the most local of scales – the home – and the close, personal and sometimes intimate connections between service providers and their clients that take place therein. In all cases the focus is on waged work. I look first at people working in the homes of others, in close daily contact with their employers, including nannies, 'dailies', domestics and other types of home-based servicing. Here ideas about the appropriateness of workers – their class attributes, their gender especially and their nationality all affect who is seen as an acceptable 'body' and so allowed into the private arena of the home. The second set of 'home' workers includes sex workers, especially those who are held against their wishes, including trafficked women forced into sex work. This chapter branches out beyond the home into massage parlours and the streets, as a large proportion of sex work takes places in public spaces, albeit often in secretive less well-surveilled spaces beyond the public gaze, as well as in the more private spaces of cars and rooms. For these workers, their body is a direct part of the exchange: they literally sell their body. And notions of desire and otherness as well as disgust are part of the exchange between seller and client. In this world, danger and violence are part of the everyday social relations of exchange. I continue this theme of embodiment and danger into chapter 6, where the workers in question also sell their bodies – in this case, their strength – as boxers, doormen and bouncers, and firefighters. In this chapter the workers in question are predominantly men, relying on stereotypical associations of men and masculinity with strength and aggression to sell their bodies in the labour market. This association raises often problematic questions about the meaning of femininity for women, including the sportswomen who also rely on bodily strength and who may feel that their skills/talent challenge their sense of themselves as 'proper' women.

The scale then shifts in part three more centrally into the larger public arena and includes what are typically regarded as more conventional workplaces, including shops and fast food outlets, hospitals, care homes and hotels. I look again, as in part two, at ways in which attributes of masculinity and femininity are mapped onto congruences between appropriate bodies for different tasks, as well as issues about race, skin colour and ethnicity in a range of types of work that involve performativity and emotional interactions between workers and clients. In some of these jobs there is a high degree of touch, as the work involves intervention on the bodies of others: these are the tasks defined as 'body work' by Wolkowitz (2006). In other jobs the contact established is less directly embodied – in shops or restaurants, for example, where the interactions are based on emotions such as empathy and deference. They fall into my definition of embodied interactive work where embodied characteristics are part of explanations of who is employed. Here the concepts of cultural imperialism and interpellation explored in this chapter are central.

In some of these chapters I draw on my own empirical research, but I also rely on case studies of service work undertaken by sociologists, geographers, economists, psychologists and others. I hope I have done justice to their arguments. In each chapter I examine who does what types of work under what circumstances, where and with what consequences. I look at how workers are assembled across space, often producing a transnational labour force, even in the most localized and spatially restricted types of work – caring for the bodies of children and the elderly. I explore questions of fantasy and desire, examining the construction of suitable workers – through interpellation and stereotypical assumptions and through the operation of the heterosexual matrix. I show how ideas about emotions, caring, servicing and the body are written into workplace social relations, into the formalized or less formal scripts written for service workers, as well as at the ways in which talk at and about work in the service sector often draws attention to, whether explicitly or not, bodies and emotions. In all these examples, there is no doubt that what is for sale in service exchanges is the bodies and emotions of the workers themselves as part of the service.

Part II
High-Touch Servicing Work in Private and Public Spaces

Plate 1 Divisions of caring labour: a mother hands her toddler to the childminder before leaving for her own waged work. Photo © Ruth Jenkinson/Getty Images.

4

Up Close and Personal: Intimate Work in the Home

A woman's work is never done.
Title of a ballad, 1629

The naturalest and first conjunction of two toward the making of a further society of continuance is of the husband and wife after a diverse sort each having care of the family: the man to get, to travail abroad, to defend: the wife, to save that which is gotten, to tarry at home to distribute that which cometh from the husband's labour for nurture of the children and family of them both, and to keep all at home neat and clean.

Sir Thomas Smith, *De Republica Anglorum*, 1583

The rest of this book is about different forms of interactive body work in the service economy. In each chapter I show how the conceptual arguments in part one aid an understanding of social relations at work in particular kinds of jobs and occupations, looking at who does what sort of work, where the work takes place and what the job involves. I explore the sets of assumptions made by employers, co-workers and clients about the appropriateness of different bodies in different spaces, as well as the differential rewards that accrue to workers in different parts of the service economy. I also examine the impact of work on the bodies of those who undertake it as well as the work that is undertaken on and for the bodies of clients and customers.

Gender Divisions of Labour and 'Work' in the Home

In this first case-study chapter, the subject is paid domestic work, that is, work undertaken for wages – typically, by women, although not always – to replace sets of tasks that at other times and in other places were undertaken within the family home 'for love'. The rise – or rather resurrection – of commodified labour within the home has been one of the most noticeable changes in the post-Fordist labour market. Despite a long history of waged

domestic service, by the mid-twentieth century a different division of labour
was common in western societies, typified by the quotation above from
Sir Thomas Smith, even though he was writing 300 years earlier. As he
explains, in sixteenth-century England it was assumed that a man's role was
to earn a living outside the home in order to provide for the needs of his
family, whereas a woman's main work was in the private sphere of the
home where, through her unwaged efforts and the sensible spending of her
husband's wages, she provided care and comfort for her children and husband. This version of family life and support dominated beliefs about the
place of women for centuries. In the twentieth century it was usually
referred to as the male breadwinner model and influenced not only wages
policies (breadwinner wages for the male working-class aristocracy and
lower 'women's' wages) but also the development of many of the institutions of the modern welfare state in the twentieth century (Lewis 1992,
1993; Daly and Rake 2003).

The extreme gender division of labour described by Smith above was
never complete. It was always an ideal rather than a reality for families where
women were forced by economic necessity to enter the labour market to earn
wages. In upper- and middle-class families, affluence brought greater freedom
to women who employed servants to undertake domestic labour. By the end
of the nineteenth century, for example, waged domestic work, often referred
to as 'being in service', provided employment for large numbers of young
working-class women, some older women and not insignificant numbers of
men. At that time, almost all middle-class families had a servant or two
who lived in the home, as well as one or two other employees, working on
a daily basis to do the heavy tasks then involved in providing an acceptable
level of cleanliness: hard and demanding work in the Victorian era. Richer
families had correspondingly larger establishments, and the aristocracy often
employed what now seems like a vast number of servants and other retain-
ers, working inside and outside the walls of the home – coachmen, valets,
housekeepers, maids and gardeners. No less than 41 per cent of all women in
waged work in 1890 were in domestic service and at the turn of the twentieth
century more Britons were employed as servants than in any other sector
apart from agriculture.

Over the first three-quarters of the twentieth century, the numbers of servants declined radically as alternative opportunities for waged work opened
up, especially for women. After the First World War, many women who had
entered previously male-dominated occupations during the war refused to
return to domestic service. A similar pattern was evident after the Second
World War, although the numbers involved by then were smaller. Although
there were still 2 million domestic workers in the UK in 1931, by 1951 their
number had shrunk to 750,000 (still 11 per cent of the female workforce) and
under 200,000 by 1961. So significant was the decline that labour market

continuity, older forms of domestic service with older

analysts assumed that domestic service was a remnant occupation, reflecting older social divisions and relations of servitude more typical of pre-capitalist societies. Sooner or later, it was argued, all women, whatever their social class and income level, would do their own housework, aided by technological improvements and new products. This, it was assumed, would result in a sort of domestic proletarianization in which all women did their 'own' housework.

However, changes in women's lives in the final quarter of the twentieth century proved these arguments wrong. Instead of disappearing, the employment of domestic servants in Britain started to rise again at the end of the 1980s and continued to do so over the last two decades. Among their number are not only female cleaners, nannies, au pairs and childcare workers, but male butlers, chauffeurs, handy men and gardeners: all replacing the work previously done until recently within the family. In July 2007, Gumtree, an internet job site, undertook a survey of domestic work in Britain (Andrews 2007), revealing that almost half of all households (48 per cent) paid for help in the home, spending on average £160 a month for six hours' labour – a grand total of £20 billion a year. Although the definition of domestic service used was a generous one, including window cleaners and dog walkers as well as nannies, cleaners and gardeners, it is clear that there has been a significant growth in commodified domestic work in recent years, fuelled, according to Gumtree, by in-migration. Later in the chapter, I look at the association between migration and domestic work. First, I explore why commodified domestic labour has increased and why it is still a predominantly feminized low-wage occupation.

Replacing women's work and commodifying care

As I explored in chapter 2, the rates of women's participation in the labour market have increased in many societies in the last twenty-five years or so. While there has been an expansion across the age range, the growth is most noticeable among younger women with children. Consequently, many of the tasks that women, especially mothers, undertake in the home on the basis of love and affection had to be reconsidered and reallocated, if women's overall workloads were not to increase to problematic levels. There seemed to be a number of options: for individual women to abandon domestic labour all together, to compress it into a shorter time and/or accept lower standards, to encourage other family members to do their share, or to replace unwaged domestic work done by members of the household with other methods of provision – either in the market for sale or by the state stepping in to assist. For a range of reasons none of these options has been entirely successful in replacing individual women's labour in the home. Empirical evidence from studies of unwaged domestic work seems to indicate a marked reluctance by men to increase their efforts. A survey produced by the Equal Opportunities Commission in 2007 reported that women still undertook 75 per cent of all unwaged work in the home, despite the protests from 'new men' about doing their share. Nor did new technologies seem to have much impact on the hours women put into keeping their homes clean and comfortable (Oakley 1974a, 1974b; Cowan 1983). It seemed then that looking for replacement services was the only answer.

The kinds of work that need to done in the home to ensure its smooth running fall into two distinct types in which the social relations and the opportunities for time compression, neglect or replacement are significantly different. The first type of work includes basic, unskilled, low-level and repetitive tasks involved in keeping the home clean: mopping, sweeping, dusting, cleaning, washing up, providing meals every single day. The second type of work is different as it encompasses the sort of affective embodied work involved in caring for others: looking after children, partners and other dependents, providing the sorts of intimate services and a loving environment in which they will flourish. These types of tasks shade into relationships based on love and affection and so are harder to conceptualize as work. This type of domestic work is much harder to commoditize than the first type as it embodies attributes of a service that are not usually recognized in classic definitions. Care, whether of children or other types of dependants, consists not only of guarding the cared-for, in the sense of making sure that no harm comes to them, but also nurturing them – loving and caring for them and ensuring that as far as possible their well-being is enhanced. Thus care is a composite good, where it is not only difficult to place a market value on the different aspects but also hard to envisage a commodified relationship that will embody all the sorts of love and care previously provided by a wife and mother. Typically, the idea of caring is bound up with notions of love and duty, with mutual reciprocity, and so it is often conceptualized as a gift relationship outside the bounds of market exchange (Titmuss 1997). Maternal love, in particular, is assumed to be 'natural', part of the social construction of femininity. Such love is both beyond value and undervalued, depending on the locus of the exchange (Folbre and Nelson 2000). Typically, caring in the home, at least when the care is undertaken by a close relation to the cared-for, is seen as beyond value and so financially unrewarded. When the exchange takes place in the market, it is undervalued, largely because of its association with the natural attributes of femininity (what is natural is not seen as a 'skill' acquired through education and training and therefore not subject to deserved financial reward). The providers of commodified care in the home – in the main women – typically are extremely poorly paid.

There is a further attribute of caring as an economic good that explains its low rewards in the market. It is difficult to achieve productivity increases in the provision of care, keeping costs high despite the poor pay for employees in this sector. Care by individuals cannot easily be replaced or substituted by alternative forms of provision. It is hard to mechanize caring or to significantly extend the scope of provision and so there is little potential for economies of scale. While it might be possible to imagine children being tagged and an alarm ringing if they crawl out of sight, this reduces the quality of care by diminishing the personal interactions based on love and affection that usually characterize these exchanges. As a consequence, most care is still provided in 'the other economy' (Donath 2000) – unwaged work in the home – as the purchase of high-quality care in the market is beyond the reach of most families. Similarly, extensive state provision, outside areas of high social need, is seen as prohibitively expensive by western governments. Nevertheless, both types of domestic provision – servicing the home and caring for others – are increasingly being provided within the cash nexus by growing numbers of cleaners, home helps, nannies and mothers' helps, often working within their employers' homes. In chapter 7, I look in more detail at caring labour undertaken outside the home, in specialist locations such as elder care facilities, hospitals and hospices. Here the focus is on the home, still considered by many to be a space of leisure and respite from the social relations of the labour market.

Location matters

Waged domestic workers by definition are employed in the homes of individual employers. Some of them live there, too. They are either the direct employees of the person purchasing their labour or sub-contracted workers, supplied by specialist agencies. In both cases there is a direct personal relationship between workers and the people for whom they variously cook, clean, shop and care for children or elderly relatives, although the responsibility for payment, tax arrangements and insurance falls on the agencies in the second case. Although the work undertaken by domestic workers is similar to all forms of low-status interactive service work, its location within the home makes it distinctive. The workplace is not a (relatively) neutral territory in the same way as a shop or a classroom. It is also the living space of the employer, a space imbued with social meaning, embodying the aspirations of its inhabitants and the ways in which they live, as well as material manifestations of relations of love and affection rather than market-based cash exchanges (McDowell 2002a; Blunt and Dowling 2006). As I explore in more detail below, these sets of meanings and assumed relationships create particular difficulties for the waged workers who have to earn their living in these same spaces. Further, the number of locations involved means that both regulating working conditions and organizing workers is particularly difficult. A great deal of domestic work is by definition unseen and invisible.

Its association with femininity means it is often regarded not as 'proper' work and so is ignored by trade union organizers. As Hondagneu-Sotelo (2001: ix) noted, based on fieldwork in the USA, 'the work of housecleaners and nanny/housekeepers constitutes a bedrock of contemporary US culture and economy, yet the work and the women who do it remain invisible and disregarded'.

Doing the dirty work: routine domestic work

The associations between dirty work, working-class women and embodiment have been explored in detail in a range of studies (Palmer 1989; Glenn 1992; Campkin and Cox 2007) in both historical (Donzelot 1980; Davidoff 1988; McClintock 1995) and contemporary analyses (Anderson 2001). McClintock, for example, argued that the home became the site of a double crisis in the nineteenth century: a crisis both of class and gender as the social relations between the social classes and between men and women were transformed by urban industrialization. In the twenty-first century, the gradual separation of the classes that was almost completed in the twentieth century is again being challenged by the re-emergence of a 'servant class', often living in close proximity to middle-class employers. Despite its recent re-emergence, however, domestic service is transformed – in terms of the labour process and for many, the labour contract. The type of work undertaken in individual homes has changed, as have the hours required to do it satisfactorily; its contractual form and organization has also altered. A new form of domestic provision as an anonymous commodified service provided by an agency has emerged as the basis of a multi-million international industry. While this industry has not (yet) superseded older forms of individualized employment contracts between employers and workers, it is an expanding part of the market.

Agency work

Domestic service provided by agencies is seldom considered in the debates about industrial restructuring in economic geography and yet the organization of house cleaning and other forms of routine domestic work is currently being restructured from a living-in occupation into a market-based commodity, increasingly supplied by firms with a national or even global reach. Domestic cleaning has become an industry, where a service rather than a body is supplied to clients. National and international cleaning services have entered the market for domestic cleaning, as well as contract cleaning for large public institutions such as schools and hospitals and for the offices of private sector firms (Ward 2004). These firms provide a service – a clean home – paid for on the basis of a time-specific contract. The employees of these firms typically are peripatetic, working on an hourly basis in many homes over the course of a week, rather than living-in and working for a single employer. The *Yellow Pages* (the British telephone directory for services) now bristles with advertisements for firms with names such as *Cinderella's, Colleen Cleaners, Daisy Dusters, Merry Maids* and *Molly Maids*, leaving the purchaser in no doubt of the gender of the service provider. The latter two providers are international franchises and their business expanded significantly in the 1990s and early 2000s (Ehrenreich 2001). Smaller, often individually owned, businesses find it harder to survive in a competitive market and a significant proportion of them go out of business each year.

Whether employed by a multinational cleaning service, a small firm or several individual employers, cleaning and other forms of domestic work are low-paid and insecure forms of employment, largely unregulated by the state. Workers typically are paid by the hour, have little security of employment and few work-related rights and benefits, such as sickness and holiday pay. The work is also socially isolating. There is often little contact either with co-workers or the customer. Workers have no status and little recognition as they tend to work when the houses they are cleaning are empty. Agency workers may work in teams but there is no guarantee that they either clean the same houses on a regular basis or that they work with the same team. Perhaps surprisingly, given that the houses they clean are usually empty, employees of the franchised firms are expected to conform to a code of conduct and a set of rules and regulations, both about behaviour in the house and about how to clean. In this way, the performance of franchise cleaning is similar to the scripted performances common in the fast food industry (Leidner 1993; see also chapter 8). As Barbara Ehrenreich (2001: 71), who worked for a short time for one of the big franchises in the USA in the late 1990s, explains:

> We have a special code of decorum. No smoking anywhere, or at least not within fifteen minutes of arrival at a house. No drinking, eating or gum chewing in a house. No cursing in a house, even if the owner is not present.

Like the domestic servants of old, many agency employees have to wear a uniform, although it is more likely to be a coloured shirt and skirt or an overall than the black and white uniforms of house servants in middle-class homes in the nineteenth century.

Cleaners and domestics – whether they are local working-class women, foreign students or recent migrants – who work for individual families on a more casual (and often cash-in-hand basis) are less likely to have either a formal code of conduct or a uniform, although they undoubtedly have a list of instructions about how to do the work. They may also establish a more personal relationship with the people for whom they clean. Some self-employed cleaners, for example, are able to build up a portfolio of 'regulars' and may work for them for several years. They may also earn somewhat higher wages than agency workers as agencies take an often-substantial slice of the wages of their employees, although as these workers have to negotiate their pay rates on an individual basis they may be underpaid in comparison to agency workers. The disadvantage for this second group of domestic workers, and for their employers, is that a personal relationship based on mutual tolerance (even respect) has to be established to keep the interaction on an even keel. By employing an agency, the end-employers (usually other women) are more able to establish an impersonal, guilt-free relationship, typically with an invisible other who cleans and mops when they are absent.

Whether self-employed or working for an agency, the tasks involved in cleaning are repetitive and tiring, often involving heavy work such as lifting or moving furniture when vacuuming, cleaning windows above head height or emptying heavy bins. Equipment may be heavy and difficult to manipulate and products used for cleaning often irritate the skin and eyes of the user and may result in damage if used over a long time period. But whatever the set of tasks involved and the equipment and the products used, because cleaning in the domestic sphere is undertaken in somebody's home and because it is associated with women's domestic duties and 'natural' inclinations, this work is extremely poorly paid. It is not interactive work in the classic sense as the customer is often not present, but it certainly is both body work and dirty work – physical effort that uses energy and demands embodied attributes of strength and repetitive movements. Hard physical scrubbing is necessary to remove the bodily traces and odours of the embodied others who live in the house, dusting to remove dirt largely composed of hair and skin cells shed by the occupants, cleaning and deodorizing to obliterate the traces left by cooking, defecating, bathing and by pets, and ironing clothes to enable middle-class service workers to perform their own work in the expected style. Cleaners get to know the intimate habits of the people for whom they work and so may come to despise, as well as, sometimes, like them, while dependent upon them to earn their living. 'This is an occupation in which the chasm of social difference plays out in physical proximity. Unlike the working poor who toil in fields and factories, domestic workers see, touch and breathe the material and emotional world of their employers' homes' (Hondagneu-Sotelo 2001: xi).

The greatest proximity and familiarity with the embodied lives of a household occurs when domestic workers live, as well as work, in their employers' homes. Workers undertaking forms of bodily care are more likely to be 'live-in' employees than those who provide house-cleaning services only. Before looking at the issues that such proximity raises, I explore this second type of domestic work – that of caring – through the lens of childcare, showing how changing expectations of motherhood have increased the demand for commoditized caring in the UK.

Doing Emotional Work: Caring for Embodied Others

When the focus turns from dirty work to caring labours, issues about the emotional world of employers become even more significant. One of the most difficult questions to answer in the analysis of the social relations embodied in caring for others as waged work in a domestic setting is the extent to which commoditized care is able to substitute for, replace or perhaps complement maternal love.

The origins of the notion of the mother as a permanently present sole carer for her own children in her own home lie in the development of the breadwinner family ideology outlined earlier, as well as in national anxieties about children's health, diet and well-being in industrial cities from about the mid-eighteenth century. With industrial urbanization and the need to regulate the growing numbers in cities, especially the working-class population, the state, as Foucault (1978) has argued, began to regulate bodies: both the individual body through mechanisms to monitor its health, its sexual practices and its usefulness, and the body of the population as a whole, what Foucault termed the 'species body', through mechanisms of control and enumeration. What emerged was a new form of state power – 'biopower' – power over bodies through discourses and practices to regulate fertility, public health, and life in the home. Foucault argued that individual women's bodies became saturated in sexuality, as well as regulated by discourses about fertility and population control that were differentiated by class position and ethnicity/racialization. As part of this regulation, feminized semi-professions such as social work and health visiting were established. Middle-class women were recruited to inspect the homes of working-class women and to provide 'advice' on sexual health, contraception and homemaking (Webster 1998).

The ideal of a 'good mother', always present in the home, regulating the lives of her children through firm but loving care, assisted where necessary by benevolent state employees, reached its apotheosis in Britain in the 1950s, building on the establishment of the postwar welfare state. In the Beveridge Report (1942), that traditional gender division of labour, recognized centuries ago by Smith, was established as official state policy and reflected in the new institutions of the postwar era. In the White Paper on Employment published in May 1944, only men were included in the category 'full' employment. Women, it was argued, had a different job: to replenish the (white) population and provide a home for their children and husbands (Lewis 1992; Webster 1998). Despite the radical nature of the new welfare state, at its heart lay an uncritical acceptance of the belief that 'the crown of a woman's life is to be a wife and a mother' (Spring Rice 1939: 95).

The good mother in the twenty-first century

This postwar version of the good mother took tenacious hold in the British psyche. Despite the rising rates of women's labour-force participation, many individual women continued to adhere to an idealized version of mothering based on caring and presence in the home (McDowell et al. 2005). This was facilitated by the development of a postwar labour market organized around part-time work for women, especially for women with children. Although part-time employment remains significant in the UK, women's growing labour market participation, especially since the 1970s, has not only required substitute care for children but has challenged the hegemonic moral commitment to remain in the home, especially when children are young. Complex questions about the extent of provision, acceptability and adequacy of substitute childcare are thus raised that remain unresolved in contemporary Britain. The state, through its insistence on workplace participation for all able-bodied adults since the mid-1990s, has an investment in establishing a different hegemonic version of motherhood. Paid work is now regarded as a moral duty; it is, as Cameron, Mooney and Moss (2002: 574) argued,

> an important component of good citizenship in an advanced liberal society, in particular for producing the ideal subject of that society: the autonomous, independent, self-regulating individual who takes responsibility for managing his/her own risks and those of their family.

All women, including mothers, with the exception of those with very young children, are expected to enter the labour market (Irwin and Bottero 2000; Lewis 2002), reshaping the boundaries between the 'private' arena of the home, the state and the market.

While femininity, domesticity and mothering used to be inextricably intertwined, breaking this relationship demands that a new set of associations with motherhood are established. A new discourse of mothering is emerging, identified by Pitt (2002) as the 'new capitalist mother'. The good mother now is a mother who enters the labour market to raise her own income and skill levels for the benefit of her children, a mother who no longer occupies the home as a continuous presence but who hands over the care of her children to another for part of the day. In the UK, active labour market policies which combine ideas about national competitiveness with policies to challenge social exclusion have been introduced (Social Exclusion Unit 1998; McDowell 2004, 2005). As well as arguing that the skills affluent middle-class women possess should not be 'wasted', poorer working-class women in deprived areas of British cities are also encouraged to seek advancement through employment, at a stroke resolving their poverty, social exclusion and the limited social capital of working-class children in these localities. In an exquisite twist, many of the new job opportunities for working-class women are in the provision of childcare for the children of the more affluent, as well as in state-provided care facilities for working-class children.

As childcare is expensive to provide, state provision is limited to the most needy families through schemes such as the Sure Start programme. All other families are expected to purchase care in the market. Instead of being a home-based 'service' provided by mothers for love, childcare in recent policies has been recast as a commodity form, reconstituted as a social responsibility enacted through the market and performed by the labour of socially unrelated others, either in the homes of individual families or in market-provided specialist facilities such as private crèches and nurseries in other locations. In the former case, the commodity relation transforms the home into a site of financially recompensed interactions, rather than a locus in which all the social relationships and interactions are assumed to be based on ties of love and affection and largely performed outside a cash nexus. In the latter, caring becomes a commodity relationship, but the spaces in which it takes place are not imbued with the same social connotations as the home, although the purchasers of care certainly hope the workers will feel affection for their charges.

In the rest of this chapter I explore caring work within the home and its consequences for the identity of both workers and their employees and the social relations between them, including a discussion of the consequences of the new mobile society identified as a key feature of the new millennium by the sociologist John Urry (2000, 2004).

Looking after other people's children

One of the most obvious differences between nannies and childminders and their employers in the UK and elsewhere is their class background. As in the nineteenth century, the working class labour in the homes of the middle class. This time, however, middle-class women are absent rather than present – working for wages themselves, unlike the 'domestic angels' of previous times. As Cameron, Mooney and Moss noted (2002: 587), in the twenty-first century 'there is a large socioeconomic gap between parents who rely on childcare and the workers themselves'. Employers of childcare typically are middle class and educated, employed in full-time high-status occupations. Their employees are poorly educated (very few British childcare workers have any qualifications above GCSE level, the basic school-leaving certificate in the UK), many of them are from working-class families, and as Gregson and Lowe (1994) found in their study of domestic work in the early 1990s, while the work was predominantly in the south of the country, many workers came from the more peripheral regions of the UK where employment opportunities were fewer. This class difference between employers and employees raises important questions about what Bourdieu (1984) termed *hexis* and *habitus* – class-specific ways of being embodied and interacting in the labour exchanges that take place within homes. A good deal of the caring work involved in looking after children is about the transmission of sets of social attitudes: about ways of talking, judgements about what may or may not be appropriate for children to read or to watch on the TV, about how to interact with others. The possession of distinctive goods and ways of living, as well as inculcating certain attitudes and beliefs, is part of the way in which the middle classes distinguish themselves from the working class (Bourdieu 1984; Skeggs 2004a). If, as I argued in chapter 3, there is a culture of class condescension in Britain at present in which the working class as a group are represented as conservative, as anti-modern, to be feared and despised as too loud, too flashy, as tasteless, eating, drinking and smoking too much in their propensity to place immediate pleasures above deferred gratification, an interesting question arises about why working-class women may be relied on to care for middle-class children. Perhaps middle-class values might be inculcated by proximity, but if the employers are largely absent when their children are being cared for, questions about surveillance and control become important. Pratt (2004: 53), in her study of live-in Filipina nannies in Vancouver, found that covert tests are set up by some employers, leaving designated books to be read to the child during the day, for example, and monitoring whether they had changed position.

The question about how to combine separate spaces and maintain surveillance of domestic workers is partially solved by new forms of technology that are able to transcend the distance between the home and middle-class workplaces. What Urry (2004: 34) termed 'inhabited machines [that] are miniaturized, privatized, digitized and mobilized' now stretch connections between home and work in previously impossible ways through mobile phones, pagers and miniaturized surveillance cameras – the logical extension of the old battery baby alarms that reassured parents of my generation that the absent child, sleeping in another room, was still secure. As Urry (2004: 35) noted in an optimistic assessment of this technology, these machines 'reorder Euclidean time-space relations, bending, stretching and compressing time-space. To inhabit such machines is to be connected to, or to be at home with, "sites" across the world – while simultaneously such sites can monitor, observe and trace each inhabited machine.'

Urry emphasized the positive effects of these technologies, suggesting, like Bauman (2000), that these machines are producing a 'liquid modernity' of interdependent flows of text, messages, people, information and images. But they are also re-cementing parent-child and parent-employee relations into structures of control and surveillance, reducing the scope for independent action by the less powerful in these relationships. The labouring bodies of the care workers, and to a lesser extent the children or elderly dependants they care for, are subjected to forms of surveillance that used to be more common in factories and prisons than within the boundaries of the home. And in homes where such forms of electronic surveillance have not been installed, nevertheless caregivers may behave as if they were watched, as Foucault argued in his definition of the mechanism of biopower, although the employers can never be sure. It seems just as likely that during the day, domestic workers engage in micro-scale strategies of everyday resistance (Scott 1985), entering prohibited spaces, breaking dietary rules or permitting children greater degrees of freedom than when their parents are present.

Despite these social controls and new forms of surveillance, a rather different rhetoric is commonly found in discussions with employers about how they think about their live-in workers, especially those providing care. In the section below, the consequences of this rhetoric – one of familialism rather than an employer-employee contractual relationship – are explored.

Relations with the 'Other': just one of the family

It is clear that the disconnection of childcare from its naturalized association with mother-love and familial care in the home, associated with complex and ambiguous feelings of duty, obligation, pleasure, desire, guilt and ambivalence, affects both the meaning of the home for women – for mothers who work for wages, for women who replace their labour – and the relationships they establish. Studies of au pairs and nannies (Romero 1992; Cox and Narula 2003) and other domestic workers who 'live in' (Radcliffe 1990; Bakuan and Stasiulis 1997) have revealed the widespread use of a rhetoric of family belonging. Through the construction of 'false kin' or quasi-family relationships, middle-class employers attempt to include waged domestic workers within relations of intimacy and caring that are seen as an ideal form of connection between carers and the cared-for. This rhetoric has an additional advantage: it acts to reduce economic obligations – to pay the basic minimum wage, for example, or to regulate hours as if childcare were a 'real' job. Steill and England (1999) in a Canadian study found, for example, a rhetoric of maternalism in the ways employers related to young, living-in domestic workers, treating them as daughters and expecting them to be available at all hours, yet these workers were often subject to rules that prevented their free use of all the spaces of the family home. Even when living conditions are excellent, nannies and other live-in carers find the restrictions of living in someone else's home impossible to ignore.

Hondagneu-Sotelo (2001: 31) reports the views of Maribel Centano, a young Latina woman in her first live-in job as a nanny in Southern California:

> I had my own room, with my own television, VCR, my private bath, and closet, and a kind of sitting room – but everything in miniature. I had privacy in that respect. But I couldn't do many things.

Her examples of feeling out of place focus on bodily presentation, food and smells:

> If I wanted to walk round in a T-shirt or just feel I was at home I couldn't do that. If I was hungry in the evening, I wouldn't come out to grab a banana because I'd have to walk through the family room, and then everybody's watching and having to smell the banana.

And so Maribel concluded:

> I could never feel at home, never.... There's always something invisible that tells you this is not your house, you just work here.

In a British study of au pairs, Cox and Narula (2003) found that the majority were not encouraged and in some cases actively prevented from using family living rooms when the family was there. Interestingly, in Britain in a 2002 Employment Tribunal ruling in which a live-in domestic worker challenged her designation as part of her employer's family, it was argued that tests such as whether the display of photos in 'public' rooms in the house included the domestic worker and whether she was included in dinner parties as an adult in her own right are good discriminators. These discursive battles over the definition and meaning of home and family (common in the US and Canada as well as in the UK) are significant as they are about the redefinition of a domestic caregiver as a worker, 'removing her from the highly gendered discursive frame of familialism and re-imagining her within the language of class' (Pratt 2004: 50). Bringing the notion of class struggle into the domestic arena transforms conventional definitions and representations of 'the home'.

Clearly, this transformation is never complete. As the growing literature about care work – both caring for children and elderly and less able dependents (Gregson and Lowe 1994; Gardiner 1997; Twigg 2000a, 2000b; Anderson 2001; Ehrenreich and Hochschild 2003; Pratt 2004) has explored, exploitation and harassment but also affection, ambivalence and guilt are common features of the relationships between employers and live-in domestic workers. In a study undertaken for the British Department of Education and Employment (DfEE) in 1999, trust in the carer where childcare was provided by an individual (nanny, childminder or relative) and a preference for someone who would show a child affection (La Valle et al. 1999: 4) were the most important factors influencing parents' choice of child carer. Clearly, co-presence and relations of obligation, affection and trust muddy the language of class struggle. The issues identified here are sensitive, difficult to approach empirically and problematic to articulate as the relations between mothers in employment and the substitute carers whom they employ are fraught with complexity and contradiction. However, a range of studies of au pairs and nannies in cities in the USA and Canada as well as in the UK have uncovered the complex and guilt-ridden set of relationships between middle-class women employers and their working-class employees, in childcare provided both within and beyond the home. Middle-class mothers may prefer to pay to have their children cared for within their own homes (Gregson and Lowe 1994) rather than in a 'public' space, but interestingly, some of the working-class young women who work as nannies or mothers' helps believe that their employers, because of their absence, are not 'good mothers'.

Other research has documented similar critical comments from employees. Cameron, Mooney and Moss (2002: 577) found that 'many [childcare workers] are opposed to the idea of working full time when their [own] children are young, subscribing strongly to the idea of "attachment pedagogy"', or as a childcare student noted 'some people put their job first, and they employ people to look after their families' (p. 579). It is self-evident that this disjunction between these young women's beliefs and the nature of their waged work must produce feelings of ambivalence. In a study of middle-class working mothers in the USA, Hochschild's (1997: 219) respondent, speaking for many other overstretched working fathers and mothers, noted: 'I am not putting my time where my values are.'

It is clear then that the embodied and emotional work undertaken as part of caring for others within the home is a particular form of embodied interactive work in which guilt, ambivalence, love, trust and obligation are all part of the social relations involved in the exchange of care for wages. And when workers live in, its location within the walls of individual homes both strengthens these feelings and makes the organization of workers as a labour force extremely difficult.

The Domestic is Global: Ethnicity, Race and Spatial Divisions of Labour

As I have argued in this chapter, cleaning and caring are associated with embodied attributes of femininity, although the former is distinguished from the latter by the absence (or perhaps more accurately the lower significance) of such intense interpersonal relationships and emotional attachment to the objects of care: a happy child is different from a clean kitchen floor. In the last part of this chapter I want to return to the differences between employers and their employees and to the simultaneous stretching and compression of relationships across space at a much larger scale than the home.

As I have already demonstrated for the UK, typically there is a significant class division between nannies, housekeepers and cleaners and their employers. In many countries, and increasingly in the UK, a further embodied distinction has long distinguished employers from their 'maids': domestic work undertaken in other people's homes 'is disproportionately performed by racialized groups' (Anderson 2000: 1). The association between ethnicity, skin colour and domestic service has a long history in the USA, where there is a marked relationship between minority status and both house-cleaning and caring jobs. The associations reflect the history of in-migration and assumptions about embodied attributes, as well as racism. As low-waged and low-status occupations, cleaning and caring work is the destination for those with least choice in the labour market. The composition of the workforce so employed varies with changes in the life chances of different groups, typically those discriminated against on the basis of ethnicity. In the USA the links between urbanization, migration and ethnicity and domestic service changed over the twentieth century as alternative employment opportunities grew. Glenn (1992) has shown how many Black women moved out of domestic service in the home into similar sorts of domestic occupations in jobs in the market, in catering and in elder care homes, for example ('public' care work is discussed in chapter 7). In the postwar era the US domestic labour force became more ethnically varied and now exhibits marked regional variations, reflecting differential patterns of settlement by older and more recent in-migrants.

> Today the colour of the hand that pushes the sponge varies from region to region: Chicanas in the South West, Caribbeans in New York, native Hawaiians in Hawaii, native whites, many of recent rural extraction, in the Midwest. (Ehrenreich 2001: 79)

In other places and at different scales, there also is a marked association between ethnicity, migration and domestic work. In European cities such as Madrid and Rome, Latin American women from Peru and Ecuador currently work as domestics (Radcliffe 1990). In countries as diverse as Italy, Taiwan and Canada, as well as in the Middle East, women from the Philippines work as caregivers and cleaners, often admitted to these countries as 'part of the family' of their employers rather than as independent workers and so not free to change their employment. This may lead to abuse, as Kalayaan, a Filipina support agency in London, has documented. Drawing on their work, Anderson (2000: 91–2) has documented the forms of exploitation suffered by domestic workers.

> The smallest things which did not please my madam resulted in abuse, shouting and slapping of my face. One dreadful occasion I washed a jumper in too hot water, this caused shrinkage. I was not only hit but almost choked to death.

Another live-in domestic cleaner reported:

> As soon as I came to London and to her house I feel like she brought me to a jail ... I have to sleep on a shelf.... So morning 4.30 to midnight I have to be up. I have no rest and I have no place to sit. She asked me not to be near the children ... she treat me as if I have bad disease.

In an interesting study of the constrained options of recent women migrants into Canada, Pratt (2004) has documented similar exploitation of women caregivers. Based on many years working with Filipina women in the Vancouver labour market, Pratt (1997, 1999, 2003, 2004) shows the ways in which migrants' origins and status not only map onto employment as domestics but also result in the establishment of a status hierarchy, both between and within domestic occupations in the city. Caring jobs – nannies and childminders – have a higher status than cleaning and housekeeping and lighter-skinned migrants are more highly valued as nannies. Drawing on a Foucauldian perspective, Pratt argues that a set of several centres of power/knowledge, operating at different spatial scales, discursively produces a construction of Filipinas as appropriate workers for low-wage caregiving occupations. These institutions include the national governments of both Canada and the Philippines – the latter exports women workers as part of its national development strategy – different levels of the Canadian state, nanny agencies and Canadian families. By constructing a migration category – live-in caregiver – the Canadian government denies new Filipina migrants citizenship rights for the first two years after entry, placing them in an inferior position to citizens in the job market, and so creating the sorts of hierarchies of eligibility based on embodied social characteristics that will become evident in other chapters. This categorization also ties women who enter under this programme to a single employer and, until the mid-1990s, excluded them from minimum wage regulations. Many Filipina women who enter Canada have professional qualifications – they are among the best educated of in-migrants to

Canada – and as the care workers interviewed by Pratt (2004: 46) explain below, the live-in caregiver scheme adversely affects their professional opportunities:

> Just because you are a contract worker, you don't have any other choice to improve or develop yourself. It takes two years before you can have an open visa here in Canada. By that time you shall have been deskilled ... two years is a long time. (Cecilia)

> After you have not worked for two years in the trade or profession that you have been trained for, you begin to doubt if you still have the ability to do your previous work. (Susan)

Despite their education and training, Filipina women find that they are restricted to the less-valued, and so often less well-paid, sectors of the home-based caring occupations. Indeed, they tend to be employed in a general capacity as housekeepers rather than nannies or childcare workers and so are expected to undertake the dirty monotonous work described earlier in this chapter. White women of European origins are preferred for the affective, emotional work of caring for children. An interview with an agency employee that placed nannies and other domestic workers made this clear:

> There are two major populations of nannies available. One is what I would call European – out of European stock – but they would be Australian and New Zealand. And there are the British-trained nanny. And several European from other descriptions. And then there are Asian, mostly Filipino. Each has different strengths. They are not the same. And they have different weaknesses. (Pratt 2004: 50)

The supposed weaknesses of Filipina applicants, according to this agency, that disqualified them as nannies included their quietness, shyness and ability to keep a house sparkling, as well as their lack of structure and discipline. And so, as Pratt agued, this duality resulted in European recruits being placed as nannies and Filipina applicants directed into housekeeping positions, showing 'the importance of understanding the intertwined cultural and social processes of identity formation and labour market segmentation' (Pratt 2004: 50) when explaining why workers with different embodied attributes are differentially ranked within the same low-status work and so differentially rewarded.

In Italy, Merrill (2006) has shown how migrant status and black skins combine to restrict African women to the least prestigious parts of domestic service. In the UK, traditionally a less diverse society than either the USA or Canada, new patterns of in-migration are also becoming reflected in the composition of the domestic workforce. Gregson and Lowe (1994), in a survey carried out in the early 1990s, found a predominantly white domestic workforce, suggesting that there is 'no close association between ethnicity, female migration and waged domestic labour ... in our study areas' (p. 123). The main distinction that they found between nannies and cleaners was on the basis of class and age. Thus, as they argued in 1994, 'the nanny in contemporary Britain is an occupational category characterized predominantly by young, unmarried women from white collar intermediate status, whereas cleaning is the domain of older, married working-class women' (p. 124). Their survey, however, was undertaken in Reading in the southeast of England and in Newcastle, the latter a distinctively white city with very low levels of in-migration or native-born minority populations. In her work less than a decade later, Anderson (2000) found significant numbers of migrants and au pairs employed by middle-class households, especially in London where the non-British born population is almost 30 per cent of the total residents.

Low-paid and low-status domestic work, especially in London, has now become a sector employing internationally mobile workers. Typically, au pairs – young women who combine childcare with learning the language – were the main source of foreign domestic labour, but this has changed as the European Union has been extended (in 2004 and 2007) and workers from Poland, Hungary and other former communist-bloc states have become a significant part of the migrant domestic labour force in Britain. While a comprehensive survey of the domestic care workers among the new A8 European population remains to be undertaken, it seems likely that these recent migrants' white skins may be a significant advantage in their search for domestic work. As I demonstrate in chapter 8 on deferential work in the hospitality sector, whiteness is an important basis of discrimination there.

Finally, while I have emphasized the lack of respect often accorded to domestic workers and the perhaps inevitably degrading nature of work that involves clearing up other people's mess, it is also indisputable that childcare is an essential, albeit woefully undervalued, interactive and subservient service occupation. Many childcare workers value their labours and often become emotionally attached to the children for whom they care. Not all employers exploit their domestic workers, although it is clear that even those who pay above the odds and provide reasonable accommodation for live-in workers benefit from the low social valuation accorded to domestic work done by women. The power of the state is too seldom used to benefit domestic workers. Their workplaces – the homes of the middle class – are seldom inspected, health regulations are ignored, maximum hours of work are often exceeded and too many migrant workers are allowed to enter advanced industrial economies under special conditions or visas that provide them with less protection than citizens. Nevertheless, as Hondagneu-Sotelo (2007) argued in the preface to a new edition of her book *Domestica*, these migrant women make a significant contribution to two societies: the one they have left behind through remittances and the one whose homes they service. And many women domestic workers also feel a pride in their work and love and respect for the people that they work and care for. However, too many women are exploited, treated as domestic servants, and subject to various forms of verbal, and sometimes physical, abuse by their employers, and left to labour in conditions of isolation. It is this feature perhaps above all others – the isolation experienced by workers whose workplace is someone else's home – that makes this form of close, embodied work so difficult to control, inspect and regulate and which still tends to lead to the assumption that it is not 'proper' work. The segregated geography of cities in the advanced industrial world and the separation of 'home' from 'work' for all but domestic workers reproduces a geography of exploitation that many labour market analysts assumed had in large part vanished by the end of the First World War.

Plate 2 Selling sex in public spaces. Photo © Peter Cade/Getty Images.

5

Selling Bodies I: Sex Work

In the [19]80s *sex work* was coined by activists in the prostitutes' movement to describe a range of commercial sex. Porn stars, erotic dancers, peep show performers, sex writers and others in the trade made it obvious that the prostitutes' movement should broaden its language. Thus the inclusive *sex worker*. But sex work is also code. In two words, 'selling sex is just another occupation'. Sex work sidesteps any pejorative meanings associated with *to prostitute* (the verb) and *prostitute* (the person).

Tracy Quan, letter to the *New York Review of Books*,
5 November 1991

If domestic service is re-emerging as a form of employment, challenging the optimistic accounts of the knowledge economy, then sex work provides a further challenge to arguments about epochal change. Here the dominant narrative is one of continuity, although as I shall show, like domestic work, a new international spatial division of labour is emerging.

Sex work is one of the clearest examples of an embodied service: the provider of the service is literally selling her or his body, and at the same time manipulates the body of the client. Selling sex is often referred to as 'servicing' a client. Bodily contact is close and personal, typically involving invasive contact and often coercion and other forms of violence. The exchange typically is a cash one, often furtive and secretive, occurring in locations ranging from open spaces, waste ground, back streets and cars, to hired rooms, hotels and specialist workspaces such as clubs and brothels. As soliciting for sex and living off immoral earnings are illegal in the UK and as earnings in the sex trade are usually undeclared and untaxed and so part of the black economy, selling sex is often seen as a distinctive form of work, involving particularly exploitative social relations. It also, as I explore in the chapter, often involves vulnerable women and men who have limited options in the labour market and who turn to sex work as their only alternative. Further, a proportion of sex workers are held against their will and forced to sell their bodies, challenging the definition of employment as the free exchange of labour power for wages.

As Tracy Quan suggests above, the term 'sex work' includes a wider range of services than physical acts, including sexual intercourse. Its introduction was a crucial move in recognizing that prostitution is neither a sin nor a pathological social characterization, nor a unique activity, but is instead a form of work that is not necessarily easy to distinguish from other forms of embodied labour. The term includes a continuum of activities from selling sex itself, through work in massage parlours and escort agencies, to all sorts of services based on the commodification and sexualization of men's and women's bodies, such as pornography and, some might argue, the advertising industry. If prostitution is defined as a form of waged work, along with these other activities, then workers surely should have the same rights to protection, economic justice and access to alternative forms of work as workers in other industries. Nevertheless, as a considerable part of the sex industry operates on the margins of legality, any investigation of its workers and their conditions of employment raises difficult questions of access and ethical issues about doing research with vulnerable workers. It is also difficult to give a definitive answer to the argument about continuity and change that is at issue here. Selling sex is not new: indeed, prostitution is often termed the 'oldest profession'. As sex work take place literally and metaphorically in liminal spaces – on the edge of legality and in transitional spaces that move between being places of work and public spaces – statistical estimates of the numbers of workers involved in the trade are hard to construct. In the Victorian era, prostitution in British cities was widely visible (Walkowitz 1980) and probably declined numerically, as well as became less public, in the twentieth century. However, if the broad definition of sex work is accepted to include the sale of embodied services in a range of jobs from pole dancing in clubs to pornographic film production, this is a second example of older forms of work becoming newly significant in a consumer economy based on the commoditization of desire and instant gratification.

The demands of the International Prostitutes Collective which represent the claims that sex work should be recognized as work are shown below. This statement was produced in response to a consultation process in 2004 about whether the UK laws relating to prostitution should be changed:

> Most people believe sex workers should not be criminalized and do not consider paying for sex an offence. Poverty and debt, major factors in driving women into prostitution, are major issues for millions of us. Yet cuts to our survival benefits and services as well as unequal pay continue side by side with billions in unrestrained military spending.
>
> The consultation paper appears to target men ("the demand"), rather than women and children ("the suppliers"), appealing to many women's dislike of the sex industry.

But given the punitive approach the UK government has adopted, following the US lead of bullying in civil liberties and criminal justice issues, women and young people can expect the worst:

- more Anti-Social Behaviour Orders landing more of us in prison;
- more deportations under the guise of cracking down on trafficking;
- higher prison sentences for women working from premises;
- military-style boot camps, with an eventual army job, for 'wayward' children as young as six.

In other words, criminalization or militarization for most of us!

It is time to stand united against our being divided between those of us labelled 'bad' and those labelled 'respectable'. No bad women, no bad children, just bad laws!

Delimiting the Sex Trade and Those Who Work In It

Despite the variety of sex work, in this chapter I focus almost entirely on prostitution and on adult women workers who typify commonly held images of sex workers. This is not to deny the existence of male, child or transgender prostitution, but rather is recognition of the severe difficulties of gaining access to these workers to collect accurate information about their working lives. I rely on the work of other researchers who have begun the difficult task of documenting the conditions under which sex workers live and work, the places where they undertake their work and the implications for their sense of self, their health and their future working lives. The discussion is based solely on studies in cities of the advanced economies, although the growing involvement of women migrant workers means that there is a global dimension to the analysis. However, it is important to recognize the specificity of the case studies that are discussed and the theoretical and policy conclusions that are drawn. Although selling sex as an occupation has certain features that unite it wherever the exchange occurs, there are significant differences in the conditions and future options of, say, a straight woman student in London working as an escort to supplement her student loan and an East European woman trafficked into the UK and forced against her will to work as a prostitute, or an Afghani teenager sold into a brothel in a city in India or Pakistan, although all of them face physical harm. What constitutes sex acts and what these actions mean are, as Foucault (1978, 1986) demonstrated, socially and culturally variable phenomena, influenced by time and place. Furthermore, the work itself, as it produces such complex and conflicting pressures on participants, has an impact on both personal and professional identity, on the possibilities of maintaining a separation between different identities and between work and non-work areas of life, as well as on the possibilities of finding alternative forms of employment. The first woman has a greater chance of being able to forget her foray into the world of commercial sex and to regard it as a transitory but perhaps regrettable form of work experience. The latter two women are trapped and may find it almost impossible to escape their life as sex workers.

As I wrote this chapter, a commercial television channel, ITV2, in the UK was running a serial starring Billie Piper, once the wholesome girl assistant to Dr Who, as a high-class tart. Called *The Secret Diary of a Call Girl*, Piper is presented as an intelligent woman in control of her life, choosing to sell her body to relatively pleasant non-violent men in upmarket hotels. The series is marketed by titillating full-size images of Piper in luxurious underwear. This image is entirely at odds with the reality of prostitution in Britain today. For obvious reasons, detailed statistics of the numbers of men and women working in the sex industry – whether broadly defined to include all the sorts of activities suggested by Quan or narrowly delimited to prostitution – are almost impossible to assemble, although a range of statistics are collected by, for example the police and welfare agencies. Reviewing Piper's television role, journalist Madeleine Bunting (2007) argued that more than half the prostitutes in the UK have been raped or severely sexually assaulted and three quarters have been physically assaulted; 95 per cent are drug users and 90 per cent want to get out. She also suggested that nearly 70 per cent meet the criteria of post-traumatic stress disorder, although she gives no source for these figures.

Most of the studies of prostitution undertaken by social scientists are based on qualitative research with particular groups. They explore a range of issues, including the ways in which sex workers make sense of their work, how and why they entered the trade, why they stay or how they manage to quit, relations with clients and the effects of the work on their non-work lives and intimate relations with friends, family and lovers. Sex work is controversial, both as a form of work and in how it is interpreted, from arguments that it is a form of work like any other to arguments that rely on notions of sin, disgrace and disease. Sex workers also disagree among themselves about its effects and impact.

Social and Spatial Divisions of Labour in Sex Work

In sections that follow, I unpack some of the social relations in sex work that differentiate workers, often ranking them in a hierarchy of desirability/acceptability depending on characteristics of the workers such as their age and skin colour in ways that parallel the ranking of domestic workers. I also look at the geography of sex work, at locations from the street to more private spaces such as massage parlours or brothels. First, however, I explore the specificity of sex work as *work* and examine the ways in which workers manage this particular form of emotional body work which is invasive, often unpleasant and may expose them to dangers. I assess the strategies adopted by sex workers to manipulate their clients and to distance themselves from the intimate relations that constitute the labour process. I consider some of the ways in which sex workers are controlled and exploited by pimps and by clients. I then return to questions of place and location, and at the end of this discussion I broaden the spatial scale by considering some of the ways in which transnational migration and trafficking is producing a global division of labour within sex work.

Sex as work

As I argued above, the notion of sex as work emphasizes the continuities between selling sex and selling other embodied dimensions of the self – social skills, technical abilities – in the workplace. This is the position of the English Collective of Prostitutes, part of the International Prostitutes' Collective, that regards sex/sexuality/the body as a commodity like any other, on sale in the market to the highest bidder. This idea 'of the commodification of the body as an object for consumption – where the legitimate leisure activity of the client is serviced by the legitimate business activity of the prostitute – has been fully embraced by certain sectors of the sex industry itself' (Brewis and Linstead 2000a: 85) and provides, in my view, a screen of false legitimacy that hides the exploitative relations involved. Relations of power are seldom equal in an exchange in which one person is selling her/his body for the gratification of another. As Sanders (2005: 321) argues, 'the ever-present occupational risks in prostitution such as violence, health-related concerns, criminalization, marginalization, exclusion from civil and labour rights and ostracism from local communities place sex work on an unequal footing in relation to the economic, social and cultural practices of the mainstream labour market.'

There is a further step, however, in the differentiation of sex work from other forms of work and that is the fact that the meaning of the sex act in different circumstances is always in contention. In a somewhat similar parallel to the arguments about women's caring work discussed in chapter 4 – which is only counted as 'work' when it is waged – sex is both a commodity exchange and a social relation based on love and affection. The meaning of the act differs depending on its context and whether or not it is enmeshed within a cash nexus. Feminist scholar Carole Pateman (1988) disagrees with the claim that selling sex is work like any other way of earning a living. There is, she argues, a difference between an employer purchasing the

labour power of workers to produce goods and services which are to be sold
for a profit and selling sexual services. The latter involves direct access to a
prostitute's body as opposed to command over how the body is used, as in
other jobs. As Pateman argues, other forms of employment do not involve
the direct use of one person's body by another, nor she suggests, are other
forms of work so clearly based on an explicit acknowledgement of the 'patriarchal right' (p. 208) of men to use others' bodies (including the bodies of
men and children). Even in sport, for example, discussed in more detail in
chapter 6, sponsors or employers command the use of a body's power, *not*
the body itself, although it might be instructive to employ Pateman's arguments when thinking through the example of the armed forces. Pateman
(1988: 207) argues that 'when a prostitute contracts out the use of her body,
she is thus selling *herself* in a very real sense' (original emphasis). Whether
this differentiation is adequate, I leave in the last instance to the assessment
of the reader. There are, however, a variety of positions within and outside
feminist scholarship that see prostitution in different ways from Pateman's
claim of distinctiveness, from the argument that it is work like any other, to
claims about empowerment in which women have some degree of choice
and control over how much they work (Perkins 1991). (For good reviews of
the alternative arguments, see Chapkis 1997; Brewis and Linstead 2000a,
2000b; Gulcar and Ilkkaracan 2002; Sanders 2005.)

Studies of sex workers themselves reveal a similar continuum of views
about the nature of the work and the extent to which those involved in it are
victims of circumstance or active participants who made a choice to engage
in sex work. Drawing on a survey of a range of work with adult women sex
workers in cities in the US, UK and Australia, Brewis and Linstead (2000b)
argue that competing academic understandings of prostitution 'are reflected
in and constituted by accounts from prostitutes themselves' (p. 171). Some
sex workers' views mirror those of Pateman: that prostitutes literally sell
themselves as part of the exchange with clients. Inga, for example, believes
that prostitution has harmed her: 'Personally it's mentally destroyed me.
I have had nervous breakdowns. I've had an overdose ... I've cut my wrists
and stuff like that' (Cockington and Marlin 1995: 169). The polar opposite
position is also found in sex workers' own accounts. Thus, some sex workers
argue that their work is work, just like any other job. Andrea, talking to
Sophie Day, an anthropologist who has studied prostitution in London for
many years, said, 'I call work, "work". That's all it is', differentiating sex
work from sex with her boyfriend as the latter involves a personal relationship (Day 2007: 35). As Day notes, women described

> how they maintained two separate realms of life, both of which involved sex,
> but of very different kinds. In brief, commercial encounters were framed by
> reference to an instrumental rationality in which services are priced, timed

and restricted to a particular workplace. Explicit calculations and negotiations were made about relative costs and benefits, and it seemed that women negotiated constantly shifting boundaries between those aspects of the sexual that could be alienated from the person and those that remained integral to a sense of self. (p. 35)

There seems to me to be a number of problems with this argument: if the boundaries are constantly shifting, as Day notes at the end of this passage, then the claim of 'two separate realms' is hard to substantiate and, if Pateman's claim about the inseparability of the body and the work holds, then it is difficult to see how sexual exchanges can be alienated from the person. Further, as I explore in a moment, for some women, coercive but 'personal' relationships with pimps, rather than ones based on a cash exchange, are a not uncommon part of sex work.

In other studies of sex work, some of the women interviewed argue that their work is a matter of choice rather than 'forced' labour by women with no other options. Thus Karen, a nude dancer, claims that 'many intelligent, self-confident women have chosen to work in this industry' (Strossen 1996: 179 and cited by Brewis and Linstead 2000b: 174); a claim supported by Nancy, who argues that 'an intelligent woman could choose a job in the sex industry and not be a victim, but instead emerge even stronger and more self-confident' (Tanenbaum 1994: 18 and Brewis and Linstead 2000b: 174). A more pragmatic position is taken by other women. Thus Mary, interviewed by O'Neill (1996: 20) notes: 'I'm a working girl ... I work with my body.' Similarly, Sara argues that sex work is a job, just work like other forms of work. She told Webb and Elms (1994: 275) if a prostitute is exploited or subjected to violence:

you have a right to go to the police, just like any other male or female in this world whether you are a prostitute or not. It's not fair the way a lot of girls get treated like dirt. They get pushed over. They are not there to be used and abused. They are there to do a job. (Cited in Brewis and Linstead 2000a: 91)

However, as I argued initially, the circumstances under which sex workers labour and the types of work they do vary considerably and have a key impact on the social relations involved, the degree of control and the extent of exploitation and personal danger, as well as (as Sara's comment reveals) how their work and they themselves are regarded by the general public. Despite these differences both in material circumstances and in the discursive representations of sex work, it seems clear that for many sex workers strategies of distancing – ways of separating the self from the task – are a key part of the commodified exchange with clients. Here Day's notion of alienation from the person seems more relevant.

Faking it: the manipulation of emotions in sex work

In commodified sexual exchanges 'it is often important for the prostitute to maintain the proper distance from the emotional demands of the client encounter and to enable the maintenance of self-identity beneath the public, professional mask' (Brewis and Linstead 2000a: 85). A range of strategies are used by sex workers to achieve this differentiation and emotional distance, as well as to cope with demands of the work. These include faking enjoyment, distancing the self from the act through drugs use, ways of reducing danger and looking out for one's self and other sex workers, and maintaining a pretence about work with family, friends, neighbours and acquaintances. As waged work is a key element in the social construction of identity for more and more people in work-dominated societies, conferring status and income as well as a sense of self-worth, this last strategy seems likely to produce severe personal conflicts, as many prostitutes regard their work with distaste and disgust and express contempt for their clients (McRae 1992).

One of the ways in which the concept of emotional labour is useful for the analysis of sex work is in its relevance to the ways in which sex workers achieve the distancing just outlined, as well as in creating the emotionally 'authentic' experience that (some) clients expect. Thus two forms of emotional labour are being undertaken concurrently: emotional labour on the self by workers to protect their integrity and the production of a version of emotional involvement with clients to satisfy their demands for an authentic sexual experience. Lisa, interviewed by Cockington and Marlin (1995: 94–5), talks about how she splits her real self from her working self:

> When I am down at work I'm not me, Lisa, any more, I'm La Toya ... you have to put on the attitude that you're that other person. Keep away from reality.... At first it was hard, splitting the two apart.

Other strategies to ensure emotional separation include controlling the time available for the encounter, limiting physical contacts and finding ways to avoid full intercourse. Sex workers in Birmingham interviewed by Sanders (2005) told her that they preferred to sell bondage and domination services rather than straightforward sex acts. Here Laura explains why:

> The domination ones, they are not actually touching your body – it's all caning them or finishing them off by hand. They are the best ones.... Those are the ones that you like because there is no body contact. (p. 327)

Similarly, in an earlier study in two US cities, Prus and Vassilakopoulos (1979), in interviews with women who sold sex in hotels, found that their interviewees also preferred as little bodily contact as possible:

> Aside from making sure that the guy is safe, that he isn't going to be roughing you up or annoying you too much, you want to make sure he doesn't touch you too much.... I don't have sexual intercourse if I can help it ... most good hookers do not do that. They do not have intercourse with a customer if they can avoid it, they do something else. (p. 59)

The interpretation of these strategies of both emotional and bodily distancing has varied. Some researchers have argued that the emotional dissonance so well described by Lisa, the attempts to compartmentalize feelings and differentiate the meaning of the sex act, result in stress, exhaustion and personal difficulties with life partners (O'Neill 2001). Further, as Lisa relies on drugs to achieve the compartmentalization, there are serious material affects on both her health and her income. Other scholars have argued that the pretence of fulfilment and the manipulation of clients' emotions allow sex workers to retain some degree of control and to speed up transactions, as well as keep their charges as high as possible (Hoigard and Finstad 1992; Sanders 2005).

It seems evident, however, that the maintenance of emotional distance, as well as the production of an adequate pretence, is not always easy to achieve. Many sex workers talk about feeling dirty or spoiled: a common feeling shared by 'dirty workers' in other forms of body work such as domiciliary care (see chapter 7). Problems may also arise when a sex worker does produce a manufactured identity and performance that fools her clients. As a street worker interviewed by McKeganey and Barnard (1996: 89) noted: 'Oh, I hate that, that really is mae [my] hate, when they try and get all lovey dovey.' And the necessary pretence is not only with clients. As Jane (interviewed by O'Neill 1996: 20) explained, 'you come home from work and your man wants to be kept happy and you've been at work all day pretending and you can't be bothered and sometimes you have to pretend with your man.'

As well as forms of pretence with sexual partners, other forms of deception are commonly practised among family and friends. As living on immoral earnings remains illegal in Britain and as the work itself involves breaching the boundaries of both what is regarded as acceptable work as well as what are conventionally defined as private practices, exchanged on the basis of love and affection, sex workers have to find ways to present a public identity that does not breach these dominant codes. Thus, their everyday life worlds, as well as their working lives, typically are constructed as an emotional performance based on pretence, often increasing stress and anxiety for those involved in this form of work. As O'Connell Davidson (1995) argues, because sex work is not a fully commodified job like any other form of work, the money that is earned does not buy the same form of respect that is typically associated with waged work. Thus, she suggests, sex workers are socially dishonoured by their work and the money that they earn. The category 'sex work', O'Connell Davidson (1996: 194) suggests, is a liminal one which exists 'in a space between two worlds; a space that is incompletely dominated by the free market ideology and incompletely detached from premarket values and codes (shame, dishonour etc.)'. This notion of two sets of values coexisting is extremely interesting and perhaps presents a challenge to the notion both of a linear progression in capitalist forms of social relations as well as a challenge to claims about the newness of the 'new' service economy. As Tilly and Tilly (1998) argued in their study of work under capitalism, old and new forms of social relations coexist both within the workplace and in the larger-scale organization of employment. Forms of informal work, for example, that were once seen as a legacy of older ways of organizing work, clearly coexist with 'modern' forms of capitalist organization in the formal economy in the cities of the advanced world. Men and women selling themselves and their labour power on street corners – whether as sex workers, building workers or domestic workers – is once again almost as common a sight in twenty-first century cities as it was in nineteenth-century cities (Davis 1990).

As well as emotional pretence with clients and ways of disguising involvement in sex work from friends and family, other forms of distancing are commonly used by sex workers. These include strategies to limit the time spent with clients and using forms of role play and routine to keep themselves detached (Browne and Minichiello 1995), and literally detaching themselves from the encounter through drug use. As I noted above, the use of both soft and hard drugs by sex workers is widespread. Brewis and Linstead (2000a: 86) suggest that 'soft drugs from caffeine and nicotine through alcohol to pills such as valium and "organic" drugs like marijuana, are widely used – sometimes to keep workers awake during long days and nights, often with nothing to do, sometimes to relax them and dull the sensation of the potential unpleasantness of the job.' Whether the use of these 'social' drugs is more widespread in the sex industry than in other forms of stressful work is hard to establish, as workers in other occupations are often more hesitant in reporting drug use to social science investigators. However, the issue of managing time and dealing with boredom is not exceptional: many other poorly paid service sector workers in monotonous and demanding work may turn to soft drugs as a way of dealing with boredom.

The use of hard drugs among sex workers is also common, although here the relationship between the characteristics of the job and drug use is harder to establish – some argue that adult women turn to sex work to support an existing habit; others that the relationship is the other way round. The Scottish women interviewed by McKeganey and Barnard (1996: 91), for example, argued that hard drugs made the work more supportable: 'if I've no had a hit, you jus' want it over an' done with. If you've had a hit, you can stand and work nae [no] bother, it doesnae bother you.' The location of the work is an important factor: many massage parlours and brothels try to eradicate drug use on the premises whereas women on the streets, whose working lives are more unpleasant, are more likely to use drugs. As Brewis and Linstead (2000a: 88) note, 'prostitution may by turns be boring, terrifying, unpredictable, disgusting and risky', but while drug use might make it more bearable, there are consequent dangers. The worker may find it harder to manage her emotions, or to produce an acceptable performance; it might be harder to insist on safe sex, for example, or workers might miss signs of imminent violence or other forms of danger. In the longer run, working to support a drug habit makes it difficult to escape sex work and move into more socially acceptable forms of work. Women who work on the streets and under the control of pimps seem more likely to use drugs than women in massage parlours, hotels or for escort agencies. In the next section I explore the different locations of sex work and the ways in which it is managed and controlled.

Exploitation and control: relations with pimps

Although sex workers are often despised or looked down on by others and the work, at best, is regarded as acting or manipulation, for many women, especially those who manage themselves and their work, the job content involves a range of skills found in many other types of work. Thus, as well as the actual erotic practices, sex work may also involve skills such as 'marketing, accounting, business planning, property management, financial control, promotion, entrepreneurship, knowledge of the law, political skills, education, acting, counselling and human resource management' (Brewis and Linstead 200b: 168). In part, the extent to which women use and acquire these work-related skills depends on the location of their work and on who exercises control over its flow.

   A common perception is that sex workers are controlled by pimps, who often exploit the women whom they control. In practice, women sex workers are employed in a range of sets of social relations and locations, although the three most common are individual women who manage their own working lives, often with or to support a partner, street workers who typically are among the youngest of sex workers and who are indeed often controlled by pimps, typically men, and indoor workers in massage parlours (or de facto brothels) whose boss is as or more likely to be a woman than a man.

While the definition of pimping is relatively straightforward – defined in
Great Britain in the Sexual Offences Act 1956 as 'living on the earnings of
a prostitute or exercising control over a prostitute' – it is clear that deciding
who is involved in pimping is more complex, as May, Harocopos and
Hough (2000: 2) explain:

> Defining pimps and pimping is complex. People are involved in the manage-
> ment of sex work in many different capacities, many of whom would not be
> generally regarded as pimping. In the first place the concept is gendered, both
> in law and in popular parlance. Pimping is almost always thought of as a male
> activity....
>
> Secondly, the popular conception of a pimp is much narrower than the legal
> definition. In specifying the offence of pimping as living on the earnings of pros-
> titution, the legislation obviously embraces a diverse range of people. In particu-
> lar, any sex worker's partner who benefits from the income of sex work is
> theoretically – and sometimes in practice – swept up by the pimping legislation.

In fact, as May, Harocopos and Hough (2000) note, in 1997 when they
began a study for the British Home Office, only 188 people were prosecuted
for pimping under sections 30 and 31 of the Sexual Offences Act in 1997.
It seems clear that the fear of being 'swept up' is hardly much of a deterrent.
It is instructive to compare this figure with the 11,459 people (almost all
women) who were cautioned or prosecuted for soliciting in the same year.
In the Sexual Offences Act 2003, further definitions of the exploitation of
prostitution were added to broaden the definition of pimping.

The concept of coercion seems to be the main way in which popular views
of pimping are distinguished from those in the Sexual Offences Act (O'Connell
Davidson 1998). Male partners of sex workers, for example – who may or
may not exercise coercive control but who may both aid their partners in their
work and indirectly benefit from it – are seldom regarded in the same light as
men who control several women as their key source of income. In their study
of the role played by pimps in women's involvement in prostitution, May,
Harocopos and Hough (2000) found that the popular view that sex workers
largely are controlled and managed by pimps was not substantiated by their
evidence. Based on 79 interviews with sex workers and pimps in four separate
geographical areas with well-defined open street sex markets, they found striking differences between street workers and workers in massage parlours. Their
key findings are outlined in the boxes on the next page. The first presents
some of their conclusions about the nature of pimping and the second shows
the differences between different forms of sex work management.

These extracts show the differences between sex workers based both on
where they work and who controls them. It seems evident that the location
of the work has a significant effect on the conditions under which workers'

The nature of pimping

1 Far from all sex workers being run by pimps and although street workers are more likely to have a pimp, a large minority – possibly even a majority – do not.
2 Many sex workers are self-managed, often supporting partners.
3 Those who are pimped are significantly at risk of physical and emotional abuse from their pimp.
4 Younger, rather than older, sex workers are more likely to be pimped.
5 Pimps play a role in drawing people into sex work, but they are not the only agents; they seem to play a larger part in locking people in, through violence and drugs for example.

Characteristics of sex work managers
Pimps

1 Pimps running street workers tend to have both criminal histories and to have 'a diverse repertoire of offending styles' of management, including the use of or threat of using guns.
2 The majority had pimped juveniles at some stage.
3 Many were heavily involved in drug dealing; most had habits and most supplied their workers to ensure compliance.

Parlour managers

1 Managers of off-street sex workers tended to be women who had little or no involvement in other forms of criminal activity.
2 Relationships between managers and workers typically were contractual rather than coercive.
3 Managers avoided recruiting juveniles and attempted to ban drug use on their premises.

Source: based on May, Harocopos and Hough (2000: vi, vii)

---

labour and in the degree to which they are subject to coercion. This is not to argue that the location of sex work is a determining influence on the nature and conditions of the work: a set of interrelated factors affect whether women work on the streets or in somewhat more salubrious locations. And although 'indoor' work was found by May, Harocopos and Hough (2000) to be preferable to working on the streets, it is clear that a different set of relations exists between pimps, trafficked migrant women and working in rooms where women are held actual or virtual prisoners. This is the more disturbing face of sex work in British cities today. If it is difficult to collect the type of information that May, Harocopos and Hough (2000) managed to compile, then it is even harder to collect accurate information about trafficking and sex work. Before I examine the relations between ethnicity, migration, trafficking and tourism and sex work, I want to turn briefly to the clients of sex workers and then discuss the boundaries between selling sex and other forms of intimacy.

Who buys sex?

As most of the research on sex work focuses on the workers rather than the purchasers of sex, less is known about the men who buy sex than about the women who sell it. Of course, clients become visible through the workers' eyes in their comments about how to avoid danger and violence and in discussions of the ways to manipulate clients, speed up transactions, and fake emotions, as discussed earlier in this chapter. However, 'men who pay for sex have been largely invisible in research, discourse and public policy' (Coy, Hovarth and Kelly 2007: 1). Although some information is available from surveys of men who attend sexual health clinics, these sorts of surveys seldom focus on the social relations of sex work. Men are usually unwilling to talk about buying sex because of the stigma surrounding prostitution and the illegality of kerb crawling and propositioning, although in a large-scale quantitative study of sexual attitudes and lifestyles (Johnson et al. 2001) one in 29 men admitted to buying sex. In London the reported incidence is higher: one in 11 men, spread across age groups, classes and ethnicities, claimed to buy sex on a regular basis (Ward, Mercer and Wellings 2005).

   To explore in more detail some of the reasons why men buy sex, Coy, Hovarth and Kelly (2007) undertook a study of male customers of sex workers in Tower Hamlets, a borough in the east of London with a high percentage of households living in conditions of multiple deprivation. Initially, the researchers planned to interview men picked up in police kerb-crawling operations or those arrested for other sexual offences. The numbers were small and perhaps not so surprisingly all the men so identified refused to be interviewed. Using a community support police officer produced six interviews and so to supplement this, advertisements were placed in a local and London-wide paper and 105 telephone interviews carried out. The extent to which the men who replied were a particular subgroup of all men who purchase sex is impossible to establish. Rather than summarize the full results of this study, I want to focus in particular on the nature of the exchange as work, and on the social and spatial relations of the exchange: how do customers behave, how is the exchange managed and controlled and where does it take place?

   Despite the research being designed to investigate the sale of street sex, the majority of the men who responded to the first advertisement tended to buy sex solely in off-street commercial locations. A further advert was placed explicitly mentioning street workers, but even then only a minority of respondents admitted to street purchase: perhaps because, suggest the researchers, the criminalization and stigma attached to street prostitution makes men less likely to volunteer to participate in research, or more likely that most prefer off-street locations (Coy, Hovarth and Kelly 2007: 5). Sex in an indoors location is more easily constructed as a 'date' whereas outdoor transactions are more clearly a 'business' exchange, often furtive, constrained by the weather and the need for vigilance. The researchers found that half their respondents (n = 69) preferred to visit a private flat, concluding that the unlicensed, clandestine nature of private flats holds an appeal for sex buyers that ' "legitimate" and/or licensed massage parlour and saunas do not' (p. 13). Most of the men interviewed were familiar with particular areas of London associated with the commercial sex industry such as Soho and Kings Cross and reported that typically they travelled beyond the area where they worked or lived to buy sex. This may be because of the greater likelihood of finding sex to buy in these areas or because they were afraid of being recognized locally, or probably a combination of both reasons.

Despite a range of reasons being given for buying sex, Coy, Hovarth and Kelly (2007) were able to categorize them into three types of discourse, which were not mutually exclusive: boasting, consuming and confessing, of which only the third was negative:

- Boasting: an equation of masculinity with sexual prowess and women's availability.
- Consuming: a leisure activity to meet a sexual 'need'; the purchase of a commodity like any other.
- Confessing: a discourse of guilt and ambivalence, some recognition of exploitation within the sex industry.

Not surprisingly, almost half (57) the interviewees regarded prostitution as 'just another job' that is 'OK if women chose to do it' (p. 16). For the boasters and the consumers the attraction of buying sex lay in its nature as a commercial transaction, with no negotiation or emotional attachment, or, for many, any respect for the workers, compared for example to their wives or partners. This led Coy, Hovarth and Kelly (2007: 22) to conclude that 'a double standard – a twenty-first century reworking of the virgin/whore dualism – persists in many men's views of women who do and do not sell sex'. However, as one of the interviewees made clear, some men regard all relationships with women as commercial transactions, however disguised in the rituals of dating. As this man explains, a commercial transaction with a sex worker might have a more 'successful' outcome in his eyes than a more conventional relationship:

> There's no questions asked, there's no crap. I could go out with a girl, take her to a bar, spend a lot of money, but now I could just give her [a sex worker] the £40 and you have half an hour with her, and you get anything you want ... there's no questions asked and that's basically it.... I've taken girls out and OK, I've take her for a meal, that cost me bloody £40.... You don't get bugger all after that.

Many of the men interviewed by Coy, Hovarth and Kelly (2007) bought sex from women of different nationalities from themselves, either from women migrants in the UK or from women during vacations or work trips outside the UK. In the next sections of this chapter, I explore some of the connections between skin colour, ethnicity, migration, travel and trafficking in the sex industry.

Foreign Bodies: The 'Exotic' Other as an Object of Desire

As I argued in chapter 3, all workers are differentially positioned by embodied characteristics including gender, ethnicity and skin colour. For women of colour, whether British citizens or more recent in-migrants, this coincidence frequently positions them as inferior, as sets of assumptions in the 'receiving' country are made about the appropriateness of different bodies for particular types of work that are culturally specific, differentially gendered and racialized, and situated within and affected by circuits of global capitalism. In Britain there is a long and distasteful history of racism that positions women and men of colour as both different from and inferior to the white majority population. Although these assumptions change over time, as do the origins and characteristics of new workers, the body of the exotic other has long exercised a fascination in western culture. The exotic otherness of, for example, Asian women – Madame Butterfly and Susie Wong – and Black women has been central in western culture, explored in art and other forms of representation and performance. Whereas the supposed fragility and femininity of Asian female bodies has been largely celebrated, Blackness has been portrayed with greater ambivalence, constructed as something to be desired and to be feared. In both representations, however, embodiment as non-white typically is associated with inferiority and so exploitation. As McClintock (1995) showed through an analysis of working women depicted in Victorian photographs, blackness was mapped onto the skins of white working-class women to construct them as different from and inferior to white and middle-class women, so justifying their exploitation, as well as evidence of both fear and delight found in Victorian middle-class men's desire for working-class women (Walkowitz 1992).

The same ambivalent construction of dreadful delight seen in many forms of representation is evident in the desire for black bodies. As numerous studies have shown (e.g. Carbado 1999; Staples 2006), the sexuality of Black men in western culture is constructed as powerful and out of control, to be both feared and envied by the white population, reflected in popular culture in images of the hypersexual 'super stud'. This social construction has a much more troubled and disgraceful history in the southern states of the USA when racial segregation and fears of miscegenation resulted in a form of apartheid. The lynching of Black men was a too common consequence of white prejudice. The bodies of Black women in the American South were more typically seen as maternal and comforting: essentialized characteristics that constructed them as appropriate domestic servants (Glenn 1992, 2001). But Black women were also objects of desire, displayed in dance halls, even in museums and exhibitions in the nineteenth and early twentieth centuries as commodified and sexualized objects of the white (typically male) gaze. Here the case of Khoikhoi Saartje Baartman, from Cape Colony South Africa, has become the exemplar. Labelled the Hottentot Venus, she was a young woman who was exhibited in London and Paris between 1810 and 1815 as a representative of deviant sexuality because of the size and shape of her body, especially her protruding buttocks. Her body evoked both primal fears of the 'Other' imagined as a sexualized savage or more primitive being, eliciting both repulsion and attraction in the male viewer. Although she died aged 26, her skeleton remained on show in a Paris museum until 1974. In 2002 her remains were finally repatriated to South Africa. A century later, in the early twentieth century, Josephine Baker, a successful Black American singer and performer, was represented in the media in similar terms as exotic, sensual and primitive. She was employed by the Folies Bergère and in the 1920s was a great success as a dancer in London, Paris, Berlin and Copenhagen. Baker, however, was also a key figure in the Harlem Renaissance of the 1920s and 1930s. Although their work is often described as primitive and sensual, it has been argued that these artists deliberately focused on primitivist examples as part of a radical critique of the white puritan ethic which they regarded as too cerebral.

In *Black Venus: Sexualized Savages, Primal Fears and Primitive Narratives in French*, Sharpley-Whiting (1999) explored the contemporary associations between black bodies, fears and attractions, arguing that the Black Venus narrative is mobilized and reasserted in moments of colonial anxiety. Magubane (2001) has built on Sharpley-Whiting's work to insist that the fear/delight narrative is better understood as 'part and parcel of larger debates about liberty, property and economic relation, rather than … simple manifestations of the universal human fascination with embodied difference' (p. 827). Sets of economic and social relationships across space result in the movement of people across space; economic inequality, poverty 'at home' and in the receiving countries, and cultural stereotypes, position female migrants as vulnerable in western econo-mies. As Westwood and Phizaklea (2000: 13) noted: 'If one asks a recently arrived migrant woman today where the opportunities for work lie in Europe, she will tell you that apart from sex work and domestic work, the avenues for employment are closed to her.' Embodied female sexuality becomes one of migrant women's key marketable assets as sex work is both an opportunity and a trap. The process of 'skin trade conscription', however, is not a matter of individual choice but part of a set of structured processes, connecting 'sending' and 'receiving' countries and involving multiple actors, recruiting, trafficking, and marketing women and, in the case of Black women, actively deploying stereotypical representations of deviant Black female sexuality as part of the service (Spanger 2002).

Jayne Ifekwunigwe (2006) has analysed the position of Nigerian migrant women sex workers in Italy, looking at the latest layer in the African diaspora that has resulted in African dispersals to Europe, in an analysis that stretches across spatial scales from transnational migration to local structures of exploitation of women's bodies in Italian cities (see also Merrill 2006). Drawing on a number of studies, Ifekwunigwe argues the transnational migratory processes that result in growing numbers of migrants from Africa in European towns and cities include 'the smuggling of West African (and North African) women and men via Morocco to southern Spain (Harding 2000); the trafficking in West African (in particu-lar Nigerian) women to Italy as part of the global sex trade (Aghatise 2002); and the strategic and "voluntary" migrations of West Africans (once again mostly Nigerian) to the Republic of Ireland' (Ifekwunigwe 2006: 208–9). She shows how the nineteenth-century colonial stereotypes of women of colour as exotic and sexually uninhibited are recycled in twenty-first century Italy in the ways in which Black sex workers are constructed as bodies for sale. However, she insists that 'the gendered, racialized and economic relations of power deeply embedded in all of these encounters and negotiations must be interpreted inside not outside the feminized cir-cuits of global capitalism and underdevelopment' (p. 213) that result in migration. Sex work is simultaneously an international question, an immi-gration issue, a legal issue, a labour issue and a gender issue; and all these perspectives must be part of its analysis, raising complex questions about research strategies and policy implications for academics and policy mak-ers. In this way, sex work becomes part of a geography of global divisions of labour rather than an exceptional activity. In this global division sex becomes a commodity in multiple ways: as an income-generating strategy in situ, as a possible passport to migration, as a step towards marriage facili-tated though both internet and actual dating agencies, and as part of the growing exploitation of women's bodies through trafficking. Like domestic workers, women sex workers are part of new global divisions of labour.

Getting a bit of the other: sex and tourism

A different aspect of the global division of labour in sex work is sex tourism in which men from western economies buy sex from women in less developed economies. Sex tourism also has a long history – from the relationships that young men established on the Grand Tour in the eighteenth century, to the mass market tourism to present-day destinations such as Thailand and the Caribbean as the network of global economic relations both expands and connects new places together. At the present time, as Brennan (2004: 156) noted in her discussion of the trade in bodies in the Dominican Republic:

> Sex tourism ... is fuelled by the fantasies of white, first-world men who exoticize dark-skinned native bodies in the developing world, where they can buy sex for cut-rate prices. The two components – racial stereotypes and the economic disparity between the developed and developing world – characterize sex-tourist destinations everywhere.

The trade typically is represented as, and is, exploitative, especially as many of the women involved are extremely young. The recent pictures of an ageing British pop star, Gary Glitter, being arrested in Thailand for preying on teenage girls is a distasteful image of what this trade involves. It is clear, too, that young boys as well as girls are vulnerable to sexual exploitation. However, as Brennan points out, exploitation is not necessarily a one-way relationship. She argues that the women whom she interviewed saw European men as potential dupes, as readily exploitable, a source of money, possibly foreign trips, even visas and marriage, although these latter prospects seldom came to fruition.

Instead of pursuing the position of men who are sex tourists, I want to discuss the development of what might seem to be a more recent phenomenon – affluent white women who also engage in sex tourism, although as the extract below from the *Observer* (a quality Sunday newspaper in the UK) makes clear, it too has a longer history:

> 'We're businessmen', says Leroy proudly. 'We sell ganja, coke and good lovin'.' His grin spreads to his eyes as he touches fists with his friend Sean. It's a traditional male Jamaican greeting expressing good wishes, friendship and respect. Sean responds, bumping his closed fist atop Leroy's. 'Respect man, to the businessmen.'
>
> It's 10 a.m. on Jamaica's breathtaking Negril beach. Bleached white sand, swaying palms and crystalline Caribbean waters stretch into the distance for seven miles. It looks endless and, on a first impression, this could be paradise.

But Negril is not as dreamlike as it looks. It is no longer visited primarily for sun, sea and sand. Instead, it is the destination of choice for an increasing number of British female sex tourists. An estimated 80,000 single women, from teenagers to grandmothers, flock to the island every year and use the services of around 200 men known as 'rent a dreads', 'rastitutes' or 'the Foreign Service' who make this resort their headquarters.

Female sex tourism is nothing new. It was reported in the late 1840s, when an Englishwoman went to Rome to take a lover. But in recent years it has grown in popularity. These days the women who participate are more likely to be single professionals than bored Shirley Valentine housewives. With females staying single longer and rising divorce rates, these holidays are expected to explode in popularity in the years ahead. Consequently they are the subject of a sudden flurry of books, films and plays examining the motivations of women who travel for sex, love and affection. (L. Martin, *Observer*, 23 July 2006)

On the screen and DVD, two recent films addressed the question of female sex tourism: *Vers le sud*, a French film based on the stories of a Canadian writer, Dany Laferriere, which stars Charlotte Rampling as a British sex-seeker in late-1970s Haiti; and *Rent-a-Rasta*, a US documentary about women tourists in Jamaica and the young Rastafarians who cater to them. In Britain the Royal Court Theatre staged *Sugar Mummies* in 2006 based on a middle-aged woman seeking love in Negril, the Jamaican beach resort described in the *Observer* article.

Despite Martin's claim in her article that up to 80,000 single women 'flock to Jamaica' it is impossible to confirm her figures. In a paper published in 2001, Jacqueline Sanchez Taylor (2001) tried to estimate the number of women involved in sex tourism in Jamaica and the Dominican Republic between 1998 and 2000. However, as she warns, figures are hard to collect, not only as tourists may be reluctant to answer questions, but also because sex tourism is so difficult to define. As the majority of analyses of sex tourism are framed in a conceptual approach that links forms of male power to the sexual exploitation of women, female sex tourism is harder to imagine and so the behaviour of first-world women who purchase 'favours' from 'local' men tends to be regarded less judgementally than that of men who buy sex. While it is clear that the exchange is a commodified one, its interpretation is complicated by an apparent reversal of typical power relationships in heterosexual encounters. As Taylor (2001: 750) notes, 'researchers often acknowledge that sexual relationships between local men and tourist women are based on an exchange of money or goods and gifts, but … they argue that the actors' narratives of romance and courtship make the term "prostitution" inappropriate.'

To explore the diversity of experiences that might be combined in the term *sex tourism*, Taylor gave questionnaires to 240 women tourists who were single or travelling without a partner in three different resorts: two in Jamaica and one in the Dominican Republic. The women were approached and asked to participate in a survey about tourism and sexual health. Perhaps surprisingly, all but 15 of them agreed. A third of these respondents had engaged in one or more sexual relationships with local men during their holiday. The women varied in age and background, the majority of them were white and many of them were visiting the resort for a repeat trip. Whereas almost all these women reported an economic element in their relationships with local men – buying meals, giving gifts and 'helping out' with cash – they did not present the encounter as a prostitute-client relationship, framing the encounters instead through terms such as holiday romances and even 'real love', although a minority admitted to a purely physical relationship. The local men with whom these women enter into relationships typically are very different from them – working-class young men, with little education and limited labour market opportunities. They are mainly hotel and bar workers or beach boys, involved in a range of economic activities as well as sex, often acting as tour guides, promoting restaurants, providing snorkelling or diving instruction, perhaps selling drugs. Their work is casualized and seasonal and inevitably low paid and their main motive for entering into sexual relationships with tourists is to gain material and economic benefits. Some women return visitors may send monthly remittances during the non-tourist season to maintain the contact across the year. Taylor argues that these men seldom identified themselves as sex workers, also constructing a discourse around mating rituals and a fantasy of courtship. In their view, only women and homosexual men can be prostitutes.

In her conclusions, Taylor challenged common preconceptions about sex work and definitions of prostitution that see it as a simple exchange of money for sex. As she argues, there is 'a wide range of different types of sexual economic relationships' (p. 760) between women and men. Furthermore, men as well as women construct their behaviour through a narrative of romance and women, as well as men, can be sexual predators, using their economic advantages in relations of unequal power. Finally, Taylor argues that women sex tourists in the Caribbean draw on the same notions of racialized otherness outlined above to justify their behaviour. As she tartly notes, 'racist ideas about black men being hypersexual and unable to control their sexuality enabled them to explain to themselves why such young and desirable men would be eager for sex with older and/or often overweight women, without having to think that their partners were interested in them only for economic reasons' (p. 760).

This study provides further support for the argument that sex work is part of a global division of labour and should be understood primarily as a form of interactive body work in which gender, class, economic inequalities and racialized otherness are an essential part of the explanation of the division of labour and the social relations of power that both construct and maintain it.

Trafficked bodies

Sex tourism and exploitation of 'Others' are not the only ways in which ethnic differences are a key part of the construction of and differentiation between sex workers. One of the most exploitative and disempowering routes into sex work for migrant women is through trafficking, although it is important to remember that not all sex workers are trafficked and neither are all trafficked men and women sex workers.

Trafficking may be defined as a process that involves a double form of exploitation – both enforced movement or coercion to leave 'home' and extreme forms of exploitation, including imprisonment, forced or bonded labour, and unwaged labour, in the receiving location. Violence is a common form of intimidation and typically workers' passports are withheld to hinder their escape. Trafficking women (and children) for sexual exploitation is thus a particular form of forced movement and forced labour in which women are recruited and held for specific purposes. It is part of a larger issue about new patterns of enforced migration, connected to new global connections and rising income differentials between countries and regions, as well as to the effects of the feminization of poverty and social and economic disruptions for many reasons, including war and famine. Women are trafficked into countries with established sex industries, often from countries where there is an indigenous sex industry. Sometimes women are part of a series of movements, not all of them forcible, from, for example, rural to urban areas in their country of origin, and from third to first world countries, sometimes with a temporary stay in transit countries (IOM 1999). This movement parallels those that Sassen (1988) noted as characterizing the movement of third world women workers to the west as 'free' workers. Whether enforced migration for sexual exploitation should be considered as an exceptional case then, or as part of a wider analysis of migration, raises complex issues. It also raises questions about the status of sex workers in the receiving country. Advocates of employment (and residence) rights for sex workers have argued that attempts to 'rescue' foreign-born women from the sex trade often result in their expulsion from the UK as illegal migrants. This illustrates the complexity of the connections between employment, migration and sex work. Trafficking for sexual exploitation can thus be seen as a moral, migration, criminal, employment, human rights or public order issue – or all of these. Furthermore, it is often seen as a gender issue. By conceptualizing the sex trade as distinguishable from other forms of trafficking, the same set of gendered assumptions about the power differentials involved in prostitution that are seen in the definitions of sex tourism are also evident here. Thus trafficking for sex has been defined as:

> Transport of women from third world countries into the EU for the purpose of sexual exploitation. (European Commission 1996: 2)

and

> Trafficking involves the transport or sale of women, with or without the consent of the victim, use of enticement, deception, force or intimidation, for the purposes of prostitution or other sexual abuse. (Kalayaan 1999)

In 2005 a European conference of sex workers was held in Brussels from which a manifesto emerged. Part of the discussion was about the relationships between sex work, trafficking and migration. The manifesto included the following claim and demand:

> The trafficking discourse obscures the issues of migrants' rights. [It] reinforces the discrimination, violence and exploitation against migrants, sex workers and migrant sex workers in particular.
> We demand that sex work is recognized as gainful employment, enabling migrants to apply for work and residence permits and that both documented and undocumented migrants be entitled to full labour rights. (Sex Workers in Europe Manifesto 2005: www.sexworkeurope.org)

In British law there is no clear distinction between trafficking, smuggling migrants and prostitution. There are no general anti-trafficking laws nor criminal legislation against enforced labour, although trafficking for sexual exploitation was recognized in the Nationality, Immigration and Asylum Act 2002 and in the Sexual Offences Act 2003. Offenders may be punished under a variety of statutes including those relating to immigration offences, labour legislation such as the new law regulating the activities of gang masters in the agricultural sector, and laws relating to pimping and soliciting. Women workers are subject to varying degrees of coercion not just in the sex trade but in domestic work, agricultural labour and the catering industry. The British government in 2007 had still not signed the Council of Europe convention on action against trafficking despite governmental condemnation of what the then-Prime Minister Tony Blair referred to at a reception to mark the bicentenary of the abolition of the slave trade as a 'modern form of slavery' (Hinsliff 2007).

Although the numbers of trafficked women in different countries and in the cities within them are difficult to estimate with any degree of accuracy, the Home Office estimated that the number of sex workers in the UK who are trapped in the doubly exploitative relations of trafficking is about 1,400 a year (Kelly and Regan 2000). It seems that many of these women may be from Eastern Europe. Estimates from the late 1990s suggest that at that time about 70 per cent of the off-street sex workers in Soho in London were of Albanian and Kosovan origins (Day 2007), although by no means all of these women had been trafficked. In their research in east London, Coy, Hovarth and Kelly (2007) explicitly asked their respondents about their awareness of trafficking and exploitation. About two thirds of the interviewees reported that they bought sex from either mainly non-British women or from British and foreign women on an equal basis. The most common answer to a question about where they thought the women were from was Eastern Europe. Although only six men actually mentioned trafficking, their view of the morality of this practice varied. The two extracts below illustrate opposing views:

> A couple of places I visited, it was Slovenian and Polish girls and I wondered whether they were kind of being forced into it. But I never discussed it with them because I went there just for sex. (p. 24)

> I don't like East European girls ... they're coerced into it, I'm not happy to think that someone's coerced. I mean I went with one and to be honest I really couldn't go through with it, because I just got the feeling that ... I felt she was sort of being compelled. (p. 23)

In 2008 the Labour government considered making the purchase of sex from trafficked women a serious offence, although it is difficult to see how it might be enforced.

Changes in the global division of labour and patterns of migration clearly affect the circumstances under which women are trafficked and the countries from which they come. As Ifekwunigwe (2006) noted, growing numbers of African women are involved in the sex trade in European cities. The number of Albanian and Kosovan women in London reflects the civil unrest there in the previous decade. Kelly and Regan (2000: 18) found in their survey of trafficking for sexual exploitation evidence of women from countries as various as Albania, Brazil, Czech Republic, Hungary, Lithuania, Portugal, Thailand and Ukraine working in the sex trade in Britain. Thus, like domestic work, sex work is increasingly part of new global divisions of labour in which women's representation is increasing, as they become a growing part both of migratory flows and of the global working class. Prostitution and sex work historically have been influenced both by older colonial histories and patterns of exploitation that produced the racialized patterns of desire, movement and inequality explored above, as well as more recently by new geographic patterns of travel, tourism and trafficking, as war, famine, economic inequality and new forms of global interconnections, facilitated by technological change, increase transnational movements. Like domestic work, the social relations of the tasks involved in sex work and its location in the more intimate spaces of the home, hotels, private rooms and, for some sex workers, in the street seem to position it as different from more formal types of service sector employment that take place in specialist locations more clearly defined as workplaces. Sex work itself is constructed through the lens of assumptions based on beliefs about femininity, naturalness and deference/power relations that influence definitions of what is private, what is personal and what is work. Both sex work and domestic work are represented as distinctive and different from 'proper' work and so considered to fall outside the frameworks of the legal regulation of employment relations.

Conclusions: Sex and Money

Sex work is a particularly complex issue for labour market theorists. As I have just argued, it challenges the basis of the distinction between the public and the private and the associated gender divisions of labour. These assumptions about what are public and what are private activities lie behind not only the structure of social provision – through the family and the state – and the separation of the public world of work from the private world of the home, but also deeply held beliefs about love, affective social relations and the distinctions between masculinity and femininity. Sex work, in parallel with commodified domestic labour, challenges the belief that women, in the main, but also other subordinate groups, are assumed to willingly offer what we might term 'relational services' – love, sex, affection, solace when weary, basic nursing care – as part of informal, privatized, naturalized and above all unpaid exchanges in the home. Once these services move into the public arena and become part of the cash nexus, these assumptions are challenged. As Zatz (1997: 277) has argued, 'what is most subversive about prostitution [is] its open challenge both to the identification of sex acts with acts of desire and to the opposition between erotic/affective activity and economic life.' Sex work, like domestic work, disturbs the too-often taken-for-granted boundaries between work, pleasure, employment and recreation that I discussed in chapter 2.

For sex workers, this subversion and transgression of boundaries produces what might be seen as a unique stigma that differentiates them from other service sector workers performing various forms of embodied interactive labour. Exchanging sex for money is still seen as morally wrong by many people. Radical feminists and the right-wing moral majority forge an awkward alliance in their condemnation of sex work, while coalitions of feminists, labour market theorists and organizations working to establish rights for sex workers as workers and citizens take a different position. Thus, different understandings of sex work depend on the particular ways in which sexuality, the family, love, affection and desire, gender, commodification, male domination, and the rights of workers and citizens, as well as the perspectives of clients and workers, are conceptualized. Sex work has been variously theorized as an autonomous labour contract/act, an exploited, subordinate, precarious or illegal labour practice, and a subordinated and exploited sex act, and/or a combination of all three.

The combination of desire, power, sex, labour and money is, of course, not unique to sex work. The premise of this book is that the social relations of employment increasingly embody all these factors in an economy dependent on the manipulation and satisfaction of desire, as Bauman (1998) so astutely recognized, and on the presentation of self by workers in performances that give the illusion of personalized interactions. These connections are exacerbated in the context of neoliberal economic and social policies that tie the rights of citizenship and income to the obligation to participate in waged work, in societies in which individual responsibilities are emphasized at the cost of models of familial or social solidarity that found more support in the mid-twentieth century, especially in the UK and Western Europe. It was a matter of belief in these former societies that sex and love are connected and that both were distinct from the realm of money, commerce, employment and competition. This belief is challenged by the widespread availability of commodified sex in post-millennium economies.

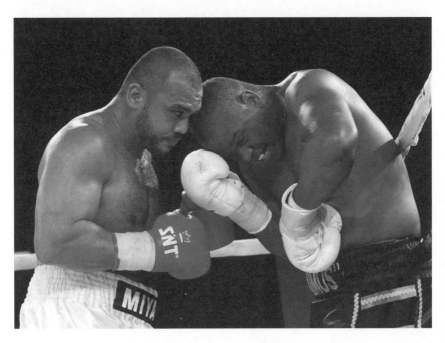

Plate 3 The 'sweet science of bruising': the embodiment of masculinity, class, and ethnicity. Photo © Duif du Toit/Gallo/Getty Images.

6

Selling Bodies II: Masculine Strength and Licensed Violence

But labour I never could abide … my fancies were set upon galleys and wars, pikes and burnished javelins, the deadly toys that bring shivers to men of ordinary mould.

Homer, *Odyssey*, c. 725 BC (translated by T. E. Shaw, 1932)

In the two previous chapters, the focus was almost entirely on women, looking at the associations between femininity and caring with domestic and sex work. For men, the main attribute associated with the masculine body at work is not its sexuality but its strength. Idealized masculine embodiment, especially when compared to a hegemonic version of fragile femininity, is constructed through its associations with physical strength. Men are tough, they stick up for themselves, they are not wimps, and in a popular advice book of some years ago, 'real men don't eat quiche'. But masculinity is also class specific – the embodied strength of working-class men is often contrasted unfavourably to the rational, cerebral and disembodied masculine virtues of middle-class men. This distinction between masculine identities and its association with particular forms of work maps onto the generic/self-programmable division (Castells 2000) outlined in chapter 2.

In this chapter, I first explore the social construction of masculinity before turning to the social and spatial relations of masculinized forms of interactive employment in consumer services, including sport, fire fighting and various forms of security provision, body guarding and door-manning. These types of work include both highly skilled and well-rewarded occupations and basic-level and poorly paid tasks in which strength is the sole requirement. Many of these jobs are high-risk occupations – the army and the fire service are obvious examples but others, especially sporting endeavours, also lead to bodily stresses and strains and so these forms of work generally provide employment for young men. Like sex work, the body itself becomes a commodity to sell, and like sex work too, some of these activities straddle a boundary – here, one between work and leisure, rather than the public and the private boundary disrupted by selling sex. For Homer's Odysseus (in the quote above), fighting was not work, but more akin to pleasure. At the bottom end of the twenty-first century labour market, the men involved in the occupations dissected in this chapter typically are working class, many of them men of colour from minority groups. Indeed, sport and the army are often escape routes for young men brought up in relatively deprived circumstances with few educational credentials and even fewer options in the feminized bottom end of service sector labour markets (Woodward 2004). Entry, however, is not open to all. Employment as a sportsman depends not only on the possession of innate skills, but also a commitment to their honing through training and other forms of investment in the body; even for the most basic level of entry into the armed services, young men – and the few women who choose the armed services as a job – must have a minimum degree of education as well as a sufficient level of fitness.

As in the previous two chapters, in exploring the case studies, I look at questions about the construction of the tasks, at the assumptions made by employers, co-workers and clients or spectators about the appropriateness of different bodies for different types of work, and the hierarchical relations that develop between workers and the sorts of financial rewards available for different types of work. I also explore the spaces and places where these forms of work are located. In each of the previous two chapters, I commented briefly on methods. Here I make methodological questions a more central part of the discussion, reflecting on issues that arise when researchers are personally involved in the work process that they are analysing. One of the most interesting aspects of the work on male embodiment and the labour process has been a growing reliance on auto-ethnography as a method of research practice. Researchers become the subjects of their own investigation through participation in the activity which they are studying. Loic Waquant's thought-provoking book *Body and Soul* (2003) describes his decision to join a boxing gym, train as a boxer and enter an official fight. After discussing some of the general implications of undertaking an auto-ethnography, I look in detail at Waquant's book and at other studies of boxing in order to raise some methodological and ethical questions about becoming personally involved in the activity being analysed. In the second part of the chapter, men working in other parts of the interactive service economy – in the night-time consumer economy and in the fire services – are the focus. Each of three case studies introduced here draws on the theoretical arguments of Bourdieu in their analysis and so provides an empirical example of the theoretical arguments of chapter 3.

Multiple Masculinities and Working Lives

There is now a large and exciting scholarship about men (see Whitehead and Barrett 2001) that, like feminist arguments, has documented the multiple ways in which men perform masculinity in the workplace and in the other spaces of everyday life. Masculinity, like femininity, is a set of practices and a position within the established gender order or the gender regime that is dominant at a particular historical period. As Connell (2001: 33) argued, gender identity is 'simultaneously a place in gender relations, the practices through which men and women engage that place in gender, and the effects of these practices in bodily experience, personality and culture'. This definition is close to that of Judith Butler's, whose arguments about gender as a performative practice were introduced in chapter 3. As Butler (1993: 94) insists:

> Rather than thinking of gender as a quasi-permanent structure, it should be thought of as the temporalized regulation of socio-symbolic norms and practices where the idea of the performative expresses both the cultural arbitrariness or 'performed' nature of gender identity and also its deep inculcation in that every performance serves to reinscribe it on the body.

Performativity is not a voluntaristic process of performance, but rather the 'forced reiteration of norms': a hegemonic ideal of compulsory heterosexuality impels and sustains gender identity.

Despite this heteronormative ideal, gender identities vary in different periods (and in different spaces). Typically, in contemporary western societies masculinity is associated with qualities such as 'control, strength, efficiency, competitiveness, toughness, coolness under pressure, logic, forcefulness, decisiveness, rationality, autonomy, self-sufficiency' (Johnson 1997: 6), although, as I argue below, different combinations of the attributes map onto diverse forms of masculinity at any one time. Men 'do gender' in different ways in particular circumstances (West and Zimmerman 1987). Further, traces of earlier sets of practices may remain. 'Any one masculinity, as a configuration of practice, is simultaneously positioned in a number of different structures of relationships, which may be following different historical trajectories' (Connell 2001: 35) and vary in different locations. Men might hold on to older versions of an authoritative rational masculinity in professional workplaces, while at home different ideas about gender relations might produce a more equitable division of labour as 'new men' embrace their emotional connections to their families. However, the workplace remains the most significant arena for the social construction of masculinity. As I argued in chapter 4, the institutions of the welfare state and the economy throughout most of the twentieth century depended on a particular gender division in which men were expected to participate in waged work.

As Connell (1995: 29) insists, 'definitions of masculinity are deeply enmeshed in the history of institutions and economic structures. Masculinity is not just an idea in the head, or a personal identity' and in capitalist societies 'the central function of masculine ideology is to motivate men to work' (p. 33). The 'Protestant ethic' originally identified by Max Weber (2002) at the beginning of the twentieth century as a key element in the industrialization of the west is part of the forced reiteration of norms identified by Butler. For men in capitalist societies, not only their incomes and their social status depend on their place in the economic structure, but their very sense of themselves as men is, in large part, constructed through the type of work that they do. However, the ways in which the connections between masculinity and workplace participation are enforced and inscribed on the body vary. Masculinity, like femininity, is a diverse set of practices and the relationships between different forms of masculine performances by differently placed bodies are structured by sets of power relations. But merely recognizing the diversity in masculinity is insufficient.

> We must also recognize the *relations* between the different kinds of masculinity: relations of alliance, dominance and subordination. These relationships are constructed through practices that exclude and include, that intimidate and exploit, and so on. There is a gender politics within masculinity. (Connell 1995: 37; original emphasis)

One of the most significant ways in which the connections between masculinity and labour market positions and practices are differentiated and structured as relationships of unequal power is through an embodied notion of class differences that maps onto manual and non-manual work. For working-class men, masculinity and masculine advantage in the labour market are based on bodily norms of strength and virility, on the ability to endure hard labour, even an insensitive toughness that permits hard bodily labour to be undertaken day after day. These attributes often lead to the development of a masculine camaraderie in the workplace that protects men at work and that strengthens male bonds, both inside and outside the workplace (McDowell and Massey 1984) and to the development of a form of hyper-masculinity (Pyke 1996) which is evident not only in workplaces but also in leisure spaces, such as working men's clubs or on the football terrace. This idealized version of masculinity is evident, for example, in studies of coal miners and fishermen, steelworkers and farm and factory labourers, as well as in the armed services, especially among the 'men' (that is, not the officer class) and it is, as is clear, a type of masculinity constructed through and reinforced by the divisions of labour in an earlier era – the years of the twentieth century dominated by manufacturing production. It is not easy to map this version of embodied masculine strength onto the interactive, generic and self-programmable jobs that now dominate in the service economy. As I show in chapter 8, this disjunction between a version of working-class masculinity and the deferential service sector jobs that are in the main the only option for young working-class men entering the labour market for the first time in the twenty-first century leads to dissatisfaction and problems of authority in the workplace. The types of highly valued work that remain open to the men who embody this working-class visceral masculinity include various forms of professional sport, the police service, the fire service and the armed services.

For middle-case men in non-manual employment, the hegemonic version of masculinity in the workplace is different. Here, bodily strength is not significant – indeed, a version of disembodied mental acuity and the capacity for rational thought is more highly valued. This cerebral version of masculinity is differentiated from both working-class male identity and the feminized bodies of women – both of which are disadvantaged by the associations between their bodies and non-rational thought. These latter bodies are out of place both in the masculinized professional work setting – the lawyers' office, the bankers' board-room, in the high-tech spaces of the IT industries – and in new media and cultural industries where creativity and masculinity are valorized. In these workspaces a version of hegemonic masculinity that is often marked by misogyny and homophobia (and sometimes racism) is evident, constructing white, middle-class men as superior to the varied cast of 'Others' who manage to gain access to these occupations. These men are not the subject here. The aim is to explore the worlds of men in working-class jobs that demand interactive embodied performances.

Men Writing About Men and Masculinity

One of the distinctive factors about academic work on masculinity and labour market ethnographies in particular is its infusion with a sense of jealousy and envy. Most of the studies of working-class men and the sorts of jobs they do have been carried out by other men – typically, middle-class academics – who may have been socially mobile (for an excellent example, see Charlesworth 2000) or whose origins are solidly middle class. These studies have a long tradition, from works such as Dennis, Henriques and Slaughter's *Coal is Our Life* (1956) onwards. They focus on traditionally male occupations in the extractive or manufacturing industries that have now largely disappeared from the UK's industrial landscape (see, for example, Samuel 1977). They include fishing, mining, and the iron and steel industries, as well as manufacturing jobs in, for example, the car industry (Beynon 1984). More recently there has been a focus on forms of interactive service work that fit notions of acceptably 'masculine' work, including professional sport. The studies in this genre, like those about education, delinquency or youth cultures, are suffused with what might be termed a 'lads studying lads' perspective in which, as Delamont (2000) has suggested, there is a strong whiff of envy from the middle-class researchers who presumably were never one of the 'bad boys', but instead high achievers at school and university.

Participant observation has been a second noticeable feature of this genre: young, and not so young, male researchers have hung out with the 'group' (Cohen 1973; Cohen and Taylor 1976), whether young workers, drug dealers, mods and rockers, car thieves or football supporters, negotiating the problems of being on the edges of semi-criminal or outright illegal activities, often with some difficulty (see, for example, Bourgois' 1995 account of avoiding the police in his work with drug dealers and users in New York City, which is discussed further in chapter 8). Clearly, this sort of methodology raises ethical issues which are magnified when, as more recently, growing numbers of researchers become full participants in the groups that they are studying. Over the last decade or so there has been a significant growth in research that involves participation rather than observation, often termed autobiographical ethnography, auto-anthropology or auto-ethnography. This move seems a logical step in the turn in the social sciences towards qualitative research methods that emphasize personal involvement, greater reflexivity and the recognition of the role of emotions in the research encounter, accepting the impossibility of making generalizable claims (Denzin and Lincoln 2000). Interestingly, however, in his assessment of this move, Anderson (2006) suggests that it is a less novel approach than many claim, arguing that there has always been an auto-ethnographic element in qualitative sociological research. He takes as an example the analyses of the sociologist Robert Park, working in Chicago before the Second World War. Park encouraged his students in interwar Chicago to study issues close to their personal lives. Where these earlier researchers differ from contemporary ethnographers, however, is in their neglect of their own values and beliefs. 'They seldom, if ever, took up the banner of explicit and reflexive self-observation' (Anderson 2006: 375), omitting their own feelings, emotions and reactions from the analysis. Indeed, sociologists, especially in the US but also in the UK, in the postwar decades adopted a position of an objective observer insisting on the scientific nature of their work, on its representativeness and its replicability by other researchers. In more recent work over the last two decades or so, influenced by postmodern arguments and the 'cultural turn' in the social sciences, it has become more common to argue that all encounters between researchers and the people they study are interactive exchanges, not reproducible by other analysts.

Critical auto-ethnography

While not dismissing the series of highly personal studies published in the social sciences and humanities between the 1970s and the early 1990s, Anderson has recently distinguished a new method: one that is not only reflexive, in which the researcher is a member of the group he or she studies, but that also retains a commitment to the theoretical analysis of broader social phenomena. He terms this current approach critical or analytical auto-ethnography, which is defined by five key features (Anderson 2006: 378):

- complete membership of the group that is the focus of examination;
- analytical reflexivity;
- making the researcher visible in the narratives produced; and yet,
- the text should include a dialogue with and understanding the points of view of the other members of the group; and finally
- a commitment to theoretical analysis of wider social structures.

Among examples of the approach Anderson advocates are Lawrence Ouellet's (1994) study of truckers, Jennifer Lois's (2003) work with search and rescue volunteers and Loic Waquant's (2003) book about boxers where the researchers undertake the forms of work that they are studying. Even so, it seems inevitable that they remain somewhat awkwardly and differently positioned from the people whom they are studying. Becoming a member of the group being analysed conveys familiarity and gives an immediacy to the research, a sense of 'being there', permitting insights into the emotional responses of the group of which the researcher is a member. Nevertheless, as the anthropologist Marilyn Strathern (1987) has argued, even though researchers may well be accepted as full members of a group, they are also social scientists, with a second or different identity to the other group members. This may lead to distraction from the embodied experiences of the others or even conflict with them, raising questions about how open to be when observing and recording others' behaviour while at the same time participating in the group interactions and work tasks. Geographer Phil Crang (1994), for example, who worked in a themed restaurant as part of his doctoral research, found it hard to both wait at tables and record the interactions between other staff and customers, and to work with, befriend and study his co-workers. There are also ethical issues about whether to be open or covert about the research. In an interesting study of women workers on a car components assembly line, Ruth Cavendish (1982) explores the conflicts she found while working on the assembly line as well as, more prosaically, noting her exhaustion after a full shift which meant she found it hard to turn to sociological analysis after a full working day. Furthermore, groups contain many different types of people and behaviours that are often difficult to differentiate between in analysis, making it difficult to produce a complete and accurate picture of the types of interactions involved in particular circumstances.

Analytical reflexivity is also not unproblematic. As researchers, social scientists typically are trained to observe the world beyond themselves and often feel uncomfortable with what may feel like a form of confessional writing. A fine line has to be drawn between an incorporation of relevant personal experiences and too great a self-absorption, the latter creating a problem in what Geertz (1988) dismissively termed 'author-saturated texts'. Further, researchers' own behaviour and motivations are also complex and multi-layered and not necessarily transparent to them: they too have complex, multiply positioned identities and are subject to ambiguity and ambivalence. Even so, Anderson's third feature of good analytical ethnography means that ethnographers must make a visible appearance in the texts they produce. Anthropological and geographical fieldworkers have long been used to writing their feelings and reactions in private notebooks, but less used to including them in the books or articles that are the eventual product of their research. An amusing and successful example of how to take care to avoid self-absorption while writing the self into the text is Renato Rosaldo's (1993) attempt to see his engagement with his father-in-law over breakfast through an anthropologist's eyes in a thought-provoking challenge to conventional notions of truth.

While not providing any guidance on how to achieve a satisfactory balance between reflexivity and self-obsession, Anderson suggests that the fourth feature of good auto-ethnography – a dialogue with others – is one way to avoid self-absorbed digressions. As Atkinson, Coffey and Delamont (2003: 57) warned, 'we must not lose sight of the ethnographic imperative that we are seeking to understand and make sense of complex social worlds of which we are only part (but a part nevertheless)'. Ethnography is above all a relational activity – interaction with others is its defining characteristic. Trying to avoid self-absorption is helped by the fifth of Anderson's defining characteristics – the aim of contributing to a theoretically located, empirically rich understanding of a set of broader social phenomena than the ones being studied. The aim ultimately is to add to, challenge or refine existing theoretical explanations of social processes, while at the same time writing a vivid work that provokes an immediate emotional response in readers but which ultimately challenges their theoretical understanding of the subject under analysis. In the next section I assess whether Loic Waquant's work on boxing meets Anderson's five criteria of analytical auto-ethnography.

Pugilism: The Body Work of Boxers

In this section I explore the boundaries of sport, leisure and work through a case study of boxing. Sporting activities are typically associated in the media and among the population at wide with men and masculinity. In previous centuries, participating in (most) sports was regarded as unlady-like. When the value of exercise was recognized, 'feminine' sports were devised to allow women's participation. From the late nineteenth century onwards, girls' schools, for example, offered netball and lacrosse to their pupils rather than rugby or soccer, seen as too combative and involving too much bodily contact (McCrone 1984). Sport as a professional activity remains a male-dominated and male-centred world, as a glance at the sports pages of any national newspaper makes clear. Contact sports, like boxing, are particularly male-dominated activities, especially at the professional level. As novelist Joyce Carol Oates (1987: 72) notes, 'Boxing is for men, and is about men, and is men.... Men who are fighting to determine their worth, that is masculinity, exclude women.' Professional sports typically valorize masculinity and male dominance, even naturalize male superiority in ways that parallel the 'natural' femininity of the caring professions (excluding medicine) discussed in the next chapter. Sport is constructed as almost self-evidently masculine:

> The institutional organization of sport embeds definite social relations: competition and hierarchy among men, exclusion or domination of women. These social relations of gender are both realized and symbolized in bodily performances. Thus men's greater sporting prowess has become a theme of backlash against feminism. It serves as symbolic proof of men's superiority and right to rule. (Connell 1995: 54)

Boxing is perhaps the quintessential masculine sport – the image of powerful sweating, bleeding, semi-naked men slugging it out in the ring seems to embody the very essence of a particular version of masculinity, even though there is a women's sport (Halbert 1997; Mennesson 2000; Lafferty and McKay 2004), represented in the film *Million Dollar Baby* (2004) directed by Clint Eastwood. Boxing as a sport and as a profession is imbued with heroic symbolism and typically comes with a set of social associations: it is dominated by working-class men, by men of colour (especially in the USA) and it is a way for men with otherwise limited opportunities to make money quickly (although most aspirants fail to become successful professionals). It self-evidently depends on an embodied version of masculine prowess and exhibitionism; it is also often regarded as a crooked sport, subject to fight fixing, illegal betting and the use of performance-enhancing substances. Boxers themselves are treated as commodities and are discarded once their performance has peaked. Perhaps above all, boxing is associated with a set of values and beliefs that is an extension of the street cultures from which fighters typically originate. Certainly the behaviour of some boxers in the recent past seems to support these contentions. The behaviour of Mike Tyson is a sobering example: his violence against women led to a jail sentence, although by no means all boxers conform to this stereotypical pattern of behaviour. In 2008 an autobiographical film of Tyson's career shown at the Cannes

Film Festival apparently embarrassed even the boxer himself, who is now free of drugs and drink and claims to regret his past misogyny and violence against women.

In a classic study of the social organization of the boxing world, Weinberg and Arond (1969) argued that boxing and delinquency have similar attributes and values – that it is almost luck and chance that directs young men's attention to one or other of these worlds, depending on whether a young man is fortunate enough to be taken up by a trainer or a role model. Masculine camaraderie, taking risks and violent behaviour characterize both worlds, and perhaps provide compensations for both social inequality and for feelings of inadequacy, although the level of athleticism, discipline and fitness demanded by boxing distinguishes it from delinquency. Vernon Scannell (1960), a British poet who worked as a boxer during his early life, has written powerfully about the compensations offered to young men through boxing. Interestingly, the feminist writer Beatrix Campbell (1993) made a similar argument about the parallels between delinquency and masculine professionalism in her book *Goliath* where she explored the roots of delinquent behaviour in what she termed Britain's dangerous spaces – local authority, city-edge housing estates plagued by troublesome youth. Her comparison, however, was between the police and the young thugs whom they attempted to control. Both groups, she suggested, found exhibitionist masculine behaviour on the streets, especially the battle of wits and the physical action involved in chases, exhilarating, although such behaviour clearly is legitimate only for the young policemen.

These arguments about class, exploitation and a form of masculinity that Connell defined as 'protest masculinity' with its origins in working-class rebellion were substantiated in an ethnographic study of a gym in a working-class inner-city area in the US by Sugden (1987). He repeated earlier claims about the connections between street and gym culture and also argued that the racism of mainstream society was paralleled in the gym, where predominantly young African-American men were treated as commodities by the mainly white promoters who pushed them into the world of professional boxing. Almost a decade later, Waquant (1992, 1995a, 1995b, 2003), a white urban sociologist, repeated Sugden's study by joining a boxing gym in inner-city Chicago (for a study in Sheffield in the UK, see Beattie 1996). I want to explore his work in detail below as Waquant reached different conclusions from Sugden, despite participating in a similar gym culture in the same sort of inner-city location. Waquant's study is particularly relevant as he develops a theoretical framework about bodily capital and embodied labour that builds on the work of Pierre Bourdieu explored in chapter 3, so his study meets Anderson's fifth criterion for analytical autoethnography – that of situating the particular case study in a wider theoretical framework.

Waquant's ethnography was carried out in an inner area of Chicago, a city blighted by poverty, deindustrialization and unemployment, where young Black men in particular faced enormous problems gaining access to employment throughout the 1980s and 1990s. Waquant (2003: 57) argued that earlier studies of boxing were based on several misconceptions. He suggested that the world of the professional boxing gym is not a parallel with the world of the inner-city ghetto in terms of its values and permitted behaviour, but rather 'the boxing gym defines itself in and through a relation of symbiotic opposition to the ghetto that surrounds and enfolds it'. He argued that there are several differences between gyms and the streets and between the reality and the image of boxing. First, he found that men who participated in the sport, as both amateurs and professionals, were not from the most disadvantaged backgrounds but were instead from the 'decent' working class; secondly, their values did not parallel street culture but instead challenged it in commitments to regular attendance, hard work, and developing a fit, healthy body rather than bodily abuse through smoking or taking illegal drugs. Thirdly, the gym provided a place of safety and security, and a degree of stability for its users in contrast to the disorganized lives of young men for whom the streets were leisure and work spaces. 'Above all, the gym protects one from the street and acts as a buffer against the insecurity of the neighbourhood and the pressures of everyday life. In the manner of a sanctuary, it offers a cosseted space, closed and reserved' for young men who are otherwise trapped by urban poverty (Waquant 2003: 15).

Unlike the young women whose lives were explored in chapter 5, boxing may offer an escape that sex work never can, although in both cases bodily abuse is common. For its members, the gym is an arena for sociability and operates on the basis of 'the unspoken code [that] members do not carry into the club their outside statuses, problems and obligations, be they work, family or love' (p. 37). Perhaps surprisingly, Wacquant also insists that 'everything takes place as if a tacit pact of non-aggression governed interpersonal relations' (p. 37). Even in sparring bouts the violence seldom escalates beyond control. Rather, violent interactions depend on a working consensus between partners based on mutual respect, and which is often playful in ways that are different from the exploitative embodied exchanges in sex work. Further, coaches regulate the mutually consented violence between boxers as it takes place in a public arena. But the gym, like other workspaces, is nevertheless structured by relations of power and status and by patterns of labour exploitation. Waquant's study largely substantiates earlier arguments about the commoditization of boxers' bodies as embodied labour power that is exploited by trainers and promoters, even though the coaches and trainers to whom he talked 'construe their work partly as a civic venture that benefits not only the boxer and his family but also the broader society as well' (Waquant 1995b: 518).

Theorizing embodiment and visceral writing

Wacquant (2003) makes a significant claim for the theoretical significance of studying embodiment and for trying to capture bodily emotions and feelings in the text. As he argues, 'the social agent is before anything else a being of flesh, nerves and senses (in the two-fold meaning of sensual and signifying)':

> Sociology must endeavour to clasp and constitute this carnal dimension of existence, which is particularly salient in the case of the boxer but is in truth shared in various degrees of visibility by all women and men ... through writing liable to capture and convey the taste and ache of action, the sound and the fury of the social world that the established approaches ... typically mute when they do not suppress them altogether. (p. vii)

Waquant draws on Bourdieu's (2000) claim that 'the social order inscribes itself in bodies through a permanent confrontation, more or less dramatic, but which always grants a large role to affectivity' (Waquant 2003: 141) (that is, the emotions and desires that lead to action). And so, Waquant argues, 'it is imperative that the sociologist submit himself [*sic*] to the fire of action in situ; that to the greatest extent possible he put his own organism, sensibility and incarnate intelligence at the epicentre of the array of material and symbolic forces that he intends to dissect' in order to understand, in Bourdieu's (2000: 141) words again, the 'relation of presence to the world, and being in the world'. Sociology must be not just *about* the body, but *from* the body, imbued with emotional connections to the subjects being investigated. For geographers, these arguments have resonance with recent work in the discipline subsumed under the label of non-representational theory (Thrift 2007), which also emphasizes affect and emotions, the importance of accepting the implications of being in the world. Some care needs to be taken, however, in considering the weight of these arguments as a research guide in different circumstances, as well as considering the practicality of adopting Anderson's five guidelines. Insisting on complete membership of a group, for example, raises complex practical and ethical issues when, for example, the group is engaged in either illegal or dangerous activities. Further, immersion is more or less possible for people with different social characteristics. Men, for example, might find the labour ward or child health clinic inaccessible or uncomfortable, and older white women would not be able to access, for example, an inner-city gang composed of minority members nor, probably, a Hell's Angels Chapter and would probably not be permitted to join a dominantly male gym.

For Waquant, becoming a member of an inner-city gym was just about possible, although, as he tells the reader, he is a skinny, white, middle-class

French man. His initial plan was to find a site where he would be able to get closer to the everyday reality of the Black American ghetto in order to combine immediacy with theory. He wanted to combine a theoretical analysis of wider social structures – an analysis of the structure and functioning of Chicago's Black ghetto and the patterns of inequality and disadvantage that were remade in the post-Fordist, post-Keynsian USA at the end of the twentieth century – and a close-up view of young Black men's lives. A friend introduced him to the gym and he was hooked, learning to box as well as keeping an ethnographic field diary for over three years. In the course of these years he managed, in his own words, 'to carve out a small place in the simultaneously fraternal and competitive world of the Sweet Science of bruising, to weave with the members of the gym relationships of mutual respect and trust' (p. x) and eventually to take part in a contest.

The text captures the physical hard work and sometimes brutal world of the sport and is suffused with the multiple points of view of the boxers, the coaches, and of Waquant himself, as the text is interleaved with long passages from his field diaries kept while he was training. Waquant not only participated in the everyday routines and training schedules of the gym, but he went to stores and the welfare office, and even cruised with the 'homies' in the housing projects in Chicago. He played pool with the men he met at the gym, went to weddings and funerals and finally witnessed the closure and demolition of the gym as part of an urban redevelopment scheme. Thus, not only are his own hopes and fears explored in the text but also the different opportunities and constraints that structure the lives of other men at the gym as their opinions and points of view are sympathetically presented by the author. The almost entirely black clientele of the gym clearly accepted Waquant's complete membership of their group: so much so that as one young man said, 'it's har' to tell you're Caucasian … d'only way one can tell you ain't black is by the way you talk an' by you bein' a Frenchman of course…. Ye, you're part of d'gym, like everyone else' (p. 11). So, even though Waquant did not originally intend to undertake an in-depth analytical auto-ethnography, his book certainly matches Anderson's five criteria for the successful achievement of this approach. But the question needs to be asked: did anyone resent Waquant's presence? If they did, did they tell him? There is no answer in the book. In the end, it is the ethnographer who chooses what to include and what to exclude and the reader is not always party to the decision.

The individual and collective body

What does the book tell us about boxing as a form of interactive body work? It clearly is a performance both in the commonsense meaning of the term and in Butler's notion of gender identity constructed within a heteronormative matrix. Is it more than this? And what are the social relations in the gym and the ring? How do boxers construct and maintain their bodily capital? What are the hierarchies of status, the patterns of inequality? The same sorts of questions that were addressed in earlier chapters are relevant here in the particular context of boxing where the relationships with the 'customers' are, in one sense, less immediate and interactive as the service providers – the boxers – are separated from the consumers by the ropes round the ring, but in another sense are more immediate, embodied and emotional as spectators shout and scream in total immersion in the spectacle of the fight. Boxing, however, depends on the social regulation of violence through a set of social relations that are constructed through and are dependent on 'intermingled affinity and antagonism' (Waquant 2003: 15–16). Drawing on Bourdieu's notion of habitus (defined in chapter 3), Waquant argues that these relationships comprise the 'pugilistic habitus'; that 'specific set of bodily and mental schemata that define the competent boxer' and which are founded on a two-fold contradiction – or what Waquant terms an antinomy:

> The first stems from the fact that boxing is an activity that seems situated at the borderline between nature and culture ... and yet requires an exceptionally complex quasi-rational management of the body. (p. 16)

The second contradiction is between the individual and the collective practices of the sport:

> Boxing is an individual sport ... it physically puts in play – and in danger – the body of the solitary fighter, whose adequate apprenticeship is quintessentially collective, especially since it presupposes a belief in the game that ... is born and persists only in and through the group that defines it. (pp. 16–17)

Boxers, through their participation in both the collective and through individual effort, become thinking and fighting machines. 'The boxer is a live gearing of the body and the mind, that erases the boundary between reason and passion ... between action and representation' and in training and fights 'transcends the antinomy between the individual and the collective that underlies accepted theories of social action' (p. 17).

Boxing also involves emotional labour in this fusion of mind and body. Although it may seem above all an example of physical embodied labour, in a small interview-based study of the role of coaches in training and promoting boxers in a British city, Sally Coates (1999) argued that in the gym there is both physical and emotional work. This is what she argues based on interviews with coaches:

> From an early stage boxers have to learn how to control and hide their emotions, particularly so in not letting their opponents know when they are hurt

physically. The experience a boxer gains in the ring is important as it expands his capacity for perception and concentration; it forces him to control his emotions in the sense of whether to repress or fuel them. The boxer may have to call forth feelings of anger but be able to control those feelings, inside and outside the ring. (p. 2)

This control and manipulation of emotions parallels the processes of management and performance found in many occupations, although boxing differs, of course, in the utility of managed anger as a professional asset. In interactive service occupations where clients and workers are involved in face-to-face interactions, anger typically has to be repressed rather than used positively. Emotions are significant in boxing in another way too: in the adrenalin rush experienced during fights and, in particular, the high that is experienced when fights are won. As an interviewee told Coates (1999: 2), 'the pay off is worth it in the end. No drugs can give you the same feeling. Even if you don't make it to the top, on the way is good too. It's the excitement, the feeling afterwards'.

Coates's study clearly does not conform to the conventions of auto-ethnography. As a woman she no doubt would have found it impossible to train with male boxers and so her findings are based on interviews with boxers and coaches rather than participant observation. Sadly, she failed to comment on the nature of the relationships that she was able to establish with the coaches and boxers to whom she talked. The reader is left to speculate on the difference that her gender made. Nor are there any details of the possible class and ethnic differences between her and the respondents. In her conclusions, however, she concurred with Waquant's positive evaluation of the role of boxing in the lives of young working-class men in deprived urban communities: 'Boxing offers an alternative lifestyle to the structural opportunities within their environment. The gym offers structure and support ... [and] the coach plays an important role in the self-development of a boxer and in building a trusting relationship' (p. 3).

Coaches, especially the head coach, are key figures in the development of the particular social practices and culture of the gym in a position that perhaps parallels the pimps and massage parlour managers who control the working lives of the sex workers discussed earlier, although the mechanisms of social control, as well as the demands of the job, are different. In the gym, social interactions, as in most workplaces, are highly ritualized and based on strict pecking orders about who talks to whom, who sits where, and who uses which equipment. The trainers and old-timers have precedence, followed by the boxers in a hierarchy of calibre and seniority. Trainers and coaches are able to make or break a professional career and they insist on strict discipline. Coaches transmit practical mastery of the corporeal, visual and mental skills needed for success through training schedules that depend on what Waquant (2003: 69) terms 'direct embodiment'. The body is a boxer's capital and it must not only be trained to react but it also has to be regulated (maintaining a fighting weight is a struggle for many men) as well as cherished and protected through the use of protective devices, bandages, ways of hardening the hands or making the chest and arms supple. Boxers use creams, lotions, vitamins and other potions and elixirs to produce a fit, supple, hard and beautiful body in the same ways as other interactive workers, including sex workers, do to conform to an idealized bodily image as well as to enhance the body's durability.

But in this world of pugilism, the boxer becomes his body in a way that differs from sex work. Moves have to become almost automatic, inscribed within the bodily schema in ways that only after endless repetition become fully intelligible to the intellect. 'There is indeed *a comprehension of the body* that goes beyond – and comes prior to – full visual and mental cognizance' (Waquant 2003: 69; original emphasis). A boxer's performance is not one of pretence, nor are the sorts of distancing techniques used by sex workers as distractions from the hard physical labour appropriate – or tolerated – for boxers. Training demands, as Oates (1987: 29) noted, 'the absolute subordination of self' in the task. Training is an essentially corporeal practice whose logic can only be grasped through repetitive actions: as Waquant (2003: 99–100) argues, 'the transmission of pugilism can only be effected in a gestural, visual, mimetic manner' that requires discipline, repetition and the gradual tolerance of greater and greater degrees of physical pain. And yet even in the mundane repetition of endless, often painful exercises, boxers are able to find some small pleasures, in glances, smiles and snatches of conversation: pleasures that were also noted by Chambliss (1989) in a study of Olympic swimmers, which in a clever juxtaposition of terms he called 'the mundanity of excellence'. Many bottom-end interactive jobs in the service sector are mundane (and some are painful), but it is incorrect to argue, as some do, that these jobs (and so the workers who do them) are contemptible, or that workers are unable to find either pleasure or self-respect in the performance of their tasks. As Hochschild (1983) argued, in the 'management of emotions' needed to produce an acceptable service performance, flight attendants nevertheless found pleasure in service. Similarly, Newman (1999) in a study of fast food workers in New York City recognized that there is scope for earning respect even in the performance of the most mundane and repetitious tasks. What successful boxers stand to gain, however, that distinguishes them from these workers, is the eventual respect of several thousands of fans as well as their fellow boxers.

As trainees, however, young boxers aim to enter a profession that is dominated by men with the least power in society. Black Americans, for example, dominated the sub-national competitions in the Midwest until 'the influx of Mexican immigrants into the lowest regions of the social space of the Midwest' (Waquant 2003: 42) from the 1980s. As Deedee, the head coach at the gym where Waquant trained, pointed out: 'if you want to know who's at d'bottom of society, all you gotta do is look at who's boxin'. Yep, Mexicans these days, they have it rougher than blacks' (p. 42). However, as Waquant noted, successful boxers are not recruited from the ranks of the most deprived members of the urban working class but from families 'struggling at the threshold of stable socioeconomic integration' (p. 43). He compared statistics on household type and income, and education levels of professional boxers at three Chicago gyms in the early 1990s and found them to be higher than the average male ghetto resident, although none of their fathers had gradu-ated from high school and all were in or had been in blue collar work. It seems too that a social position and boxing success are correlated with success, as men from slightly 'better' backgrounds seem to be more success-ful. In the transition from amateur to professional status, there is 'a better chance of being successful if the fighter can rely on a family environment and social background endowed with a minimum of stability' (p. 53). Overall, as the owner of a gym in Detroit argued, 'most of my boys, contrary to what people think, are not that poor' (p. 44). They are not, in the main, from the disorganized 'dangerous class' in the inner city.

Beatrix Campbell (1993) in her book *Goliath* mentioned earlier, used the term 'dangerous' in her subtitle, although she linked it to place rather than class, reflecting the spatial patterns of segregation on both US and UK cities that mean the urban poor and working classes live in identifiable areas, segregated from their more affluent co-residents. In Britain the 'dangerous class' of men includes the white working class as well as minority group members, as likely to live on outer authority housing estates as in the inner-city areas where US ghettos typically are found. In chapter 8 I look at the working lives of deprived and disadvantaged young men from poor working-class families in both Britain and the USA, who cannot access the world of sport and who have little option but to undertake feminized work in the retail, fast food and other sectors of the service economy. This type of work is in conflict rather than congruent with the versions of protest masculinity common among inner-city youth and so fails to provide an outlet for mas-culine aggression and frustration and a pathway, for the few, to success and relative affluence in the way that boxing is able to do. In the final part of this chapter, however, I continue to explore the links between masculinity and embodied aggression in a range of other jobs for working-class men.

Doormen, Bouncers and Other Workers in Britain's Night-Time Economy

Waquant (2003) argued that boxers become habituated to a higher degree of pain than 'ordinary' men find tolerable. Through training, modification of their bodily schema and the uses to which the body usually is put through adherence to a strict physical regime, the body of a boxer becomes 'an intelligent and creative machine capable of self-regulation' (p. 95). To opponents of the 'manly art' this may seem an exaggeration, although there is no doubt that (many/ some) professional boxers produce regulated and controlled performances that depend on a great deal more than raw masculine strength and force. I want to look next at bouncers and doormen to see whether similar forms of controlled aggression are part of the performance of workplace tasks or whether this form of work is simply a legitimized, but relatively unregulated, form of work for thugs.

One of the less pleasant aspects of the expansion of a consumer-dominated economy based on the commodification and instantaneous satisfaction of desire has been the significant growth of a night-time economy, largely marketed on the basis of pleasure-seeking opportunities and typically fuelled by the consumption of copious quantities of alcohol (Finney 2004). Many of the centres of British towns and cities become public leisure arenas, especially on Friday and Saturday evenings. Other cities, both in the UK and in countries in mainland Europe, attract the dubious benefits of 'stag night' and latterly 'hen night' tourism that turns them into party venues for the relatively affluent youth of Europe. For the bouncers and doormen (as well as the police) who are employed to regulate the night-time economy, their bodies are both the tool of their work and the target of their opponents. In this sense their work tasks parallel those of boxers. But there are also significant differences. For door staff (usually men) and bouncers, violence is not the total sum of their work and is often avoidable by adopting other forms of evasive tactics. It is also often unexpected, rather than a necessary part of the work. Further, for bouncers, job-related violence must remain within the law: overstepping both legal and socially sanctioned forms of social interaction is not permitted. This creates a grey area in which the employees may be uncertain of how to behave – in, for example, dealing with women who they decide should be restrained. Unlike the interaction between a boxer and his opponent, the rules and regulations that determine face-to-face interactions between pleasure seekers and bouncers are not refereed. As the point of contact for customers hoping to enter a venue, bouncers and door staff have the entire responsibility for making and enforcing decisions. In the UK contractors who supply employees to undertake different types of security work should be regulated under the Security Industry Authority (SIA), although at the end of 2008 the Scheme remained voluntary.

Access to work of this type – typically, low paid and low status – depends in large part on the embodied characteristics of a version of working-class masculinity. Sizeable men, with a threatening appearance, able to intimidate, preferably by looks if not through violent actions, have an advantage in controlling and disciplining the often excited, unruly and sometimes intoxicated bodies of pleasure seekers, beginning or ending a night out in the clubs, pubs and other venues that make up a key part of Britain's booming night-time economy (Winlow 2001). These associations, however, confirm the status of this work, which is often seen as little more than licensed thuggery. The work is riddled with conflicts and tensions, not only because of its immediate requirements and the need to control undisciplined crowds of people – paradoxically through violence to imbue civility – but also located in the wider-scale contradiction between the expansion of economic activities based on excessive consumption and the principles of desire and hedonism with an occupation that depends on surveillance, regulation and, when necessary, violence (Hobbs et al. 2002; Monaghan 2002). As Monaghan (2002: 406) describes below, many of the urban pubs and clubs that are expanding in British cities are predominantly hedonistic arenas of bodily display and pleasure to which door staff regulate access:

> Hedonistic and highly sexualized nightspots – populated and constituted by strangers whose decorative bodies are cultivated for gendered displays rather than verbal communication – provide the sensual and spectacular contexts where the door supervisors' 'bodily capital' is transformed into an economic resource. (p. 406)

The sensual and spectacular environment is, of course, the indoor arena, whose warmth and crowded physicality is in stark contrast to the often-cold outdoor arena where pleasure seekers queue and shiver in flimsy clothes that are a marked contrast to the often semi-military, dark and substantial clothing worn by the bouncers. The primary role of those who regulate access into spaces of pleasure is to assess, through a quick bodily inspection, whether to admit or refuse entry to putative customers. Conflict and confrontation, often expressed through violence, is a typical accompaniment of this 'filtering' work and emotions often run high, although door staff typically aim to exercise control through non-violent mechanisms and self-restraint is a valued attribute in employees. The rules of regulation that are operated include judgements about clothing, including appropriate footwear (the 'no jeans, no sports shoes' rule of some clubs), possession of alcoholic drinks, the level of inebriation and the general attitudes of those attempting to enter. Refusal often offends and leads to bodily contact between customers and door staff.

Is an auto-ethnography of this type of employment possible?

Doing research on the social relations that take place between door staff and potential clients is not straightforward. For Waquant, it was possible to become a participant in boxing in an overt way as a leisure pursuit. To become a full participant in the working life of bouncers necessitates

obtaining a job, and ideally becoming registered as a Licensed Premise
Supervisor: a legal requirement which is not always enforced. Even presenting oneself as a potential employee is not necessarily easy for an academic researcher. Gender plays a part in acceptability – there are few
women in the job – and size and appearance, as I noted above, are prereq-
uisites that are not always easy to meet for typically desk-bound workers.
Applying for and taking a job with no intention of staying raises ethical
issues, although some researchers claim that using a covert methodology
is acceptable (Calvey 2000; Winlow 2001), and actually carrying it out
may involve emotional, professional and legal and bodily dangers. Lee
Monaghan, in his work on bouncers, became a participant observer rather
than a full participant and so was unable to meet the requirements of a
full auto-ethnography discussed above. He fulfilled many of the attributes
necessary to fit the typical occupational requirements of corporeality, a
certain physical appearance and an intimidating bodily presence:

> I know my male gender, relative youth [he was 30 in 2000], and bodily capital
> [muscular, weighing approximately 16 stones and 6 feet tall] are [key]
> resources.... Although I possess a non-violent self-image, my embodied social
> history consisting of lifting weights and boxing ... have rendered me willing
> and able to work as a doorman. (Monaghan 2002: 409)

Interestingly, although he told the men he was observing that he was an
academic researcher, because of his physical appearance, they ignored the
differences and 'primarily treated me as a working doorman' (p. 410). His
brawny appearance clearly cancelled out his more cerebral attributes.

Monaghan worked at several different establishments during the course
of his study, although all of them catered for a young, typically white and
heterosexual clientele. The door staff at these sites were also young (aged
between 19 and 45, with a preponderance of younger workers), working
class, almost all men (only 5 out of the 60 door staff Monaghan observed
were women), and in the main white (50 out of 60). His study was undertaken in a southwest city in England with a small ethnic minority population: elsewhere, men of colour often find the associations between skin colour
and perceived threat works in their favour. Many door staff combined this
work with another occupation, often a full-time one in the formal economy.
Other jobs included milk delivery, scaffolder, fire fighter, karate instructor,
and perhaps more surprisingly, a trainee accountant, internet consultant and
an aircraft engineer. The job also attracted students and recent graduates
who found it hard to find more permanent work in more prestigious occupations. Indeed, for many bouncers, their work was informal rather than
part of the formal economy. Workers seldom had a contract, there was no
official entitlement to benefits such as holidays or sick pay, and wages were

frequently paid often 'off the books', although at rates that typically were above the national minimum wage. Monaghan began his study in 1997 and it was published in 2002. He gives a figure of £7.50 an hour for the average rate of pay, although it is not clear over what period the average was constructed – but even so the hourly rate was considerably higher than the official minimum wage in 2002. The degree of informality and off-the-books recruitment and payment make it a particularly difficult occupation to regulate.

For the workers, the industry combines body work, violence, physical danger, male camaraderie and often sexually predatory behaviour in ways that construct it as an almost exclusively masculine domain. Although this is clearly interactive service work, its social conditions are more reminiscent of the type of male bonding found in manufacturing industries in earlier decades (McDowell and Massey 1984). It differs, however, from industries such as coal, iron and steel, or fishing in its close interactions with clients. Door staff and bouncers regulate bodies, often in an explicit hands-on way. As Monaghan notes, 'for at least some door staff, licensed premises were seductive and captivating "outlets" where masculine affirming violence could be realized' as well as sites providing the 'opportunity to meet attractive, sexually available women' (p. 411).

Most male door staff work on the boundary of public and private space – between the street and the inside of the premises, whether a club, pub, dance hall or other space of pleasure, paralleling the liminal spaces of prostitution, although door staff have a legitimate presence denied to sex workers. They patrol an ambiguous boundary – the doorway – one that many customers assume they have the right to cross. Typically, there is a status hierarchy – although one that is not necessarily reflected in differential pay rates – between the head doorman, his co-workers on the door and lower-status workers who patrol the inside of the building. Ironically, these 'floating' inside workers have a greater degree of autonomy, patrolling different areas of the space depending on the temporal rhythms of the event, the day of the week and the size of the 'audience'. While these indoor floaters may have a greater degree of autonomy, they also find parts of their job difficult. For men, regulating women's often drunk and sometimes partially clothed bodies in the semi-privacy of the women's toilets challenges bodily norms and boundaries, and is seen as a problematic part of the job (Hobbs, O'Brien and Westmarland 2007) and so leaves scope for women to find a place as workers in this dominantly male world. As well as in circumstances when customers are physically vulnerable, whether only temporarily as in the toilets of a club, or more permanently in sites such as hospitals and clinics where men and women are in intimate or vulnerable positions, ill and perhaps undressed, gender becomes important in ways that challenge the stereotypical version of a security worker. Thus in a Canadian study of security workers, a male guard noted: 'It would not be nice for a man to walk into a woman's hospital' and a women security worker reported uneasily: 'I wouldn't feel right going into the men's change room' (Erikson, Albanese and Drakulic 2000: 308).

For the door staff, decisions about which spaces are appropriate to enter seldom arise as they are, by contrast to the indoor workers, more restricted spatially. Monaghan (2002) notes, however, that they do leave their post for various reasons, including, on occasion, for covert sexual encounters with willing customers in 'backstage' areas of the venue: an opportunity that is seen by some doormen as a perk of the job. Others reported their pleasure in engaging in a degree of violence, but distinguished what they regarded as 'good violence' from inappropriate violence. The former was a level of physical intimidation that was sufficient to repel a customer who had been refused entry without endangering the doorman himself, drawing the attention of the police and resulting in a complaint being laid against him. The habitual and widespread use of CCTV in British leisure venues and city centres is now an important method of surveillance which regulates the regulators' behaviour.

Although the security industry as a whole, and door work in particular, is an industry dominated by men, a small number of women find employment here (Hobbs, O'Brien and Westmarland 2007; O'Brien, Hobbs and Westmarland 2007). However, as Erikson, Albanese and Drakulic (2000: 294) have argued, the industry provides a good example of what they term 'resegregation'. This typically occurs when occupations which have been dominated by one sex – men in the case of bouncers and door staff – start to recruit the other sex. However, when women are the new entrants what often happens is that they are allocated to what are seen as appropriately female tasks – women's work that needs a feminine touch – checking the drunken women in the toilets, for example. As a consequence these parts of the job are redefined as 'women's work'. When men are the new entrants – into, say, primary school teaching or nursing – more typically they are singled out as suitable for promotion and their progress through the ranks of responsibility often is more rapid than women's progress in the same jobs (see chapter 7). Lisa Adkins (2003, 2005) has termed this resegregation 'reterritorialization' in her critique of the work of scholars such as Beck, Giddens and Urry that was introduced in chapter 3. As I noted there, in contrast to claims about new forms of mobility and a loosening of the traditional constraints of class and gender, Adkins argues that gender differences are reinscribed in the emerging divisions of labour evident in the new millennium.

Security firms provide willing bodies for other sorts of surveillance and regulation: to ensure the safety, for example, of both people and buildings, protecting the cars of politicians, guarding goods in warehouses or art in museums, and transporting prisoners to court. In all cases these are low-level and poorly paid jobs, for which those with little choice in the labour market but with physical stamina to exchange for wages are the main body of recruits. Non-British born men are often a significant part of these labour forces. As I noted above, many of the workers in the security industry are employed informally and so are hard to regulate and even those who have theoretically been vetted by the Security Industry Authority (SIA) are not always legal employees. As a whole, the security industry has a poor reputation for checking the credentials of its employees; periodically, scandals occur when it emerges that ex-criminals or illegal workers have been employed in sensitive positions. In November 2007, for example, the Home Office admitted that 11,000 of the 40,000 non-EU migrants who had been licensed by the SIA to work as private security guards had no right to work in the UK. Many among these 11,000 employees were working in sensitive posts, guarding Whitehall departments, for example, on Metropolitan Police contracts (Travis 2007) and so placed right at the heart of the British state.

The discovery of so many irregular and illegal workers reveals the precarious nature of this type of bottom-end employment and, often, the desperation of some of the men doing this sort of work. For most employees, most of the time, the work is routine, boring, unskilled, pays poorly and is low in social status, as well, as on occasion, dangerous. For women in the industry, because they are less likely to be obviously physically aggressive, sometimes more interesting opportunities might open up. However, most women tend to find themselves restricted to specialist niches – where the client is a woman for example, or in undercover work where a woman might be less visible. Providing or advising on in-house security, especially in private households, as well as some forms of private investigation work are areas where women are beginning to make an impact, as well as in more traditional areas such as shop-floor walkers and in-store detectives. When women do undertake front-line security work, then gender stereotypes also operate. Women are seen as less violent that men and so less likely to exacerbate a difficult situation. Thus the assumption that women prefer to minimize rather than exaggerate conflict becomes a job qualification:

> The girls [*sic*] are a bit smarter. They are going to listen to what we say in the training about violence and stay away from it, follow the rules and regulations ... [and] accomplish the same thing. (Erikson, Albanese and Drakulic 2000: 307)

Despite this recognition that a woman's presence and perhaps her preference results in defusing difficult situations, masculine strength, size and readiness to fight remain the key job attributes in this type of work. As I shall show in chapter 8, defusing potentially awkward situations is also constructed as an appropriate part of a woman's job when working in the hospitality industry.

Risking the Body: Other Forms of Dangerous Masculine Body Work

While boxing and bouncing are both jobs that predominantly rely on embodied masculine strength and the associated dangers, they also reproduce male camaraderie and provide elements of fun. Mellor and Shilling (1997), for example, have suggested that the pugnacious interactions between male door staff and male customers are not necessarily conflictual but may produce 'collective effervescence and an embodied sensual solidarity' (p. 20). Waquant's account of the boxing gym is suffused with examples of male bonding. In other jobs constructed around ideals of masculine embodiment, danger is the predominant feature. Here high-risk jobs – becoming a soldier or a fire fighter – necessitate some acceptance that the risk of death is an everyday part of the job, although for soldiers typically only during periods of active combat. Clearly, risks are enormously enhanced when, as at present in the USA and UK, soldiers are involved in action overseas. But fire fighters regularly risk their lives. These jobs too, like boxing, are clearly part of the service sector. They involve the sale of bodily strength as part of the service provided, but do not parallel interactive forms of work in which there is a direct exchange with the customers. The 'customer' buying the bodily efforts of the armed forces is hard to determine – the nation-state for whom soldiers fight are employers rather than customers, whereas the invaded states are perhaps reluctant consumers of the service. The clients of the fire service are rather easier to identify: they are the owners and/or occupiers of the properties where fires have broken out and on many occasions contacts are indeed interactive and embodied as occupants are physically rescued from dangerous situations.

One of the most interesting questions to ask about these forms of typically masculine work (although in all cases women are now recruited in small numbers) is what is the motivation for seeking such high-risk work? Are these (mainly) young working-class men whose employment options are limited? Do they become used to taking risks on a daily basis or is this an attraction of the job, providing the emotional 'highs' identified by boxers? And what are the social relations at work that ensure their obedience to orders in high-stress circumstances? In Bourdieu's terminology, what is the specific habitus of the services? What beliefs and emotions construct their working worlds and persuade them to sell their labour power in ways that put them at risk of bodily mutilation, even death? These questions are not easy to answer and often the men themselves who have taken these forms of work have not directly addressed them. This is made clear by the growing numbers of soldiers who are currently leaving the armed forces as the daily losses of life in Afghanistan and Iraq are an insistent reminder of the dangers of their working lives.

While perhaps not as obviously dangerous as soldiering at a time of war, working for the fire service is a demanding job. The service is what Coser (1974) termed a 'greedy institution' — workplaces where conventional divisions between work and home, employment and leisure, friends/family and co-workers typically are blurred and where the demands made on workers' time are excessive. Fire fighters and policemen, like doctors, are, for example, expected to be readily available in emergencies and to return to their place of work or sites of fires and accidents on demand.

Fighting fires

In this last section of chapter 6, I explore the motivations of men who risk their lives at work, drawing on a study by Matthew Desmond (2007) of wildfire fighters in the Forest Service in a rural part of northern Arizona in the USA. I have chosen this paper, despite it being an example from a rural rather than an urban area as the other case studies are, because Desmond also uses Bourdieu's notion of habitus, explained in chapter 3. He demonstrates that the primary habitus of young men in a small settlement in rural America is transformed into a more specific habitus that acts to unite wildfire fighters. By primary habitus, Desmond means the family and class backgrounds of these men that predispose them to take risks, fitting them for the rigours of fire fighting. The study is also an appropriate comparator as Desmond, like Waquant and Monaghan, bases his conclusions on an ethnographic study undertaken when he himself served as a wildfire fighter while he was a doctoral student. These ethnographic studies of dangerous occupations seem to appeal to a particular type of male researcher. Here is Desmond's (2007: 392) description of his research methods:

> By taking the 'participant' in 'participant observation' seriously, by offering up my body day and night, to the practices, rituals, and thoughts of the crew, I gained insights into the universe of fire fighting, insights I gleaned when I bent my back to thrust a pulaski (a specialist furrowing tool) into the dirt during a direct assault on a fire or when I moved my fingers through new warm ash to dig for hot spots. My body became a field note, for in order to comprehend the contours of the fire fighting habitus as deeply as possible, I had to feel it growing inside me.

This, I believe, awkwardly positions the reader as a mere spectator, or worse, a voyeur: too scared, too old or the wrong sex to be able to undertake a study like this one. I found myself both envious and irritated as I read on.

The 14 wildfire fighters at the centre of Desmond's study are seasonal workers. They take temporary employment during the cold months and then in the summer move to live in forest camps to be both ready and close to when fires break out. The job is dirty, dangerous and unglamorous.

Yet, nevertheless, in Desmond's opening paragraphs where he describes what the work involves, there is a sense of excitement and a feeling of men's heroic efforts against the elements:

> When a blaze bursts, fire fighters rush off to the scene armed only with hand tools, flame-resistant clothing, hard hats and fire shelters to 'dig line' in front of a lethal and combustive force that has no purpose other than to destroy.

He continues:

> Those who chose to square off with the 'Black Ghost' must regularly work 14 (or more) hours on end, crawling through ash and dirt, hiking through steep terrain carrying twenty pounds of gear, swinging axes and shovels, sometimes miles away from the nearest paved road, let alone the nearest hospital. And they don't always win. (Desmond 2007: 388)

Sometimes the fire rages out of control and, in the worst circumstances people are killed: local residents as well as fire fighters.

What is it that persuades these men to risk their lives? While not for an instant denying their bravery, part of the clue perhaps lies in the tone of Desmond's text. This work may be dirty and dangerous, yet it is seductive. As participants in extreme sports report, danger is/may be exhilarating. And fire fighting, as Waquant suggested about boxing, is a carnal activity, a visceral experience: the risks of approaching a fire, in summing up the danger of intense heat, are made at the level of the body and cannot be fully translated into verbal accounts. Even so, as Desmond insists, an explanation of who is prepared to become a fire fighter needs a more complex analysis within a more satisfying theoretical framework which includes an assessment of the effects of structural categories such as race and class, in combination with Bourdieu's concept of habitus. This latter idea captures the shift from calculation to practice, perhaps from mind to body, in its emphasis on bodily knowledge rather than mental calculations. It is a dispositional theory of action and so appropriate for understanding the forms of embodied and interactive work.

So what bodily dispositions, what ways of thinking about being and acting in the world, do young men who are prepared to risk their lives as fire fighters exhibit? What features of their lives during childhood and adolescence predispose them to take up this work? Perhaps unsurprisingly, it is the combination of being young, masculine, rural (a combination of their familiarity with the great outdoors: these young men know the area around and despise 'city folk', and the lack of alternative well-paid job opportunities) and their working-class background that influences their decisions. Men who enlist in the ranks of the armed forces in both North America and the

UK (Woodward 2004; Woodward and Winter 2007) come from similar backgrounds, although ethnicity plays a key part too, especially in the USA, but also in the UK (Dandeker and Mason 2001). And like men from small settlements where there is little alternative work or dominated by a single industry such as coal, boys and young men follow their fathers into the same sorts of dangerous masculine work.

For the young men with whom Desmond worked and talked, binary categories – urban/rural; indoors/outdoors – were a key distinction in the development of their sense of self identity and their primary habitus. These men not only lived outdoor lives but rather despised the soft, indoors types who had little connection to the countryside. But they also distinguished themselves from environmentalists and firmly aligned themselves with the aims of the Forest Service. The binary distinction between the country and the city thus maps onto an opposition between government-sponsored forestry and environmentalism. 'Arguments over where the fault of a devastating fire season lies, how best to manage forests, the politics of logging and thinning, the treatment of endangered species, and hunting and camping rights are all manifestations of a power struggle between independent environmentalist groups and governmental organizations, such as the US Forest Service' (Desmond 2007: 402). The former see the latter as advocating invasive management techniques; the latter regard the former as misinformed middle-class zealots, despite often having little or no contact with environmentalists.

While accepting the view of the Forest Service is part of the development of the specific habitus of fire fighters, acclimatizing themselves to actually fighting a fire demands the development of other competencies, including loyalty, team work, the ability to operate the pumps and hose, and to fell trees. But Desmond suggested it is more than this. Talking of how he managed to work in harmony with the 'country boys' from Arizona, he argues as follows:

> We knew the language of fire fighting, so to speak, because we shared a linguistic disposition formed (and informed) by a shared country-masculine history. Because we possessed a similar history, we also possessed a common code that allowed us to communicate meaningfully and seamlessly....
>
> [We] adjusted our bodily movements to one another.... Again this was possible because we shared a country-masculine history that predisposed us to such actions.... When my country-masculine habitus encountered itself in the postures, movements, rhythms, gestures, and orientations of my crewmembers, it recognized something familiar, something known deep down, and, accordingly, it synchronized with other manifestations of itself, creating a chemistry of sorts that coordinated action. (p. 477)

This passage worries me. Despite my sympathies with Desmond's approach, it seems too close to that naturalized and essentialized version of a masculinity that has operated for too long to exclude the 'Other': people of colour, women, urban weaklings. Although undisclosed, I assumed the 'country boys' were all white. I also wondered what they made of Desmond who, despite his claims of similarity, was at the time a graduate student at an elite university. We are left wondering at the end of the paper.

Conclusions: Men, Bodies and Danger

This chapter has combined the study of dangerous work with an assessment of the practices of ethnography and auto-ethnography where the researcher participates in the lives of his (in this case the male pronoun is accurate) research subjects. Like the earlier examples of feminized body work, interactive work and embodied labouring, most of the jobs discussed here are low status, low paid, often boring and, like sex work, dangerous. Here too, close and intimate contacts with the bodies of clients and sometimes co-workers is part of the labour process. And the occupations considered are largely filled by men (and some women) with few educational credentials or skills other than their masculine strength and their willingness to suffer as qualifications to exchange for income.

The focus was on detailed ethnographic studies and the methodological questions they raise, and so the wider statistical picture of the size, nature and characteristics of the workers in the different occupations has been neglected in this chapter. Clearly, there are many more bouncers than boxers, and probably more fire fighters and soldiers, but as both boxing and security work is often a casual form of work, sometimes more of a hobby than a job, and often undertaken in combination with other forms of work on a casual or part-time basis, it is difficult to know exactly how many and who work in these industries. What is certain is that for many men in unskilled forms of body work in the private sector, their employment contracts are uncertain and their attachment to the labour market precarious. For fire fighters, soldiers and police men and women, public service brings greater security and for some better financial rewards. But in all these professions, employment is precarious as the danger involved in the work brings uncertainty about how long the body might be able to endure and survive the necessary risks.

The methodological emphasis of this chapter also raises some unanswered questions and I hope provides an example of how to think critically about methodology when reading a case study. These studies are fascinating – their immediacy and attention to details grabs the readers' attention and draws them into the text. It seems a particularly appropriate way of capturing the embodied nature of the work – the sweat, dirt, hard physical effort – as well as generating awe and admiration (as well as irritation) at the balls of the researcher for getting so involved. But it is important to ask what might have been excluded – is, on the one hand, close involvement conducive to attention to detail, or, on the other hand, to developing an overview of the world participated in? Does Desmond's insistence on the significance of the past to explain current patterns of work produce a backward looking view, even a romanticized version of a job that cannot admit to the need for change? Is family, gender and class background – the primary habitus – always an essential element in explaining how individuals enter particular jobs and adapt to its cultural practices, as Desmond seems to assert? Is this an appropriate/necessary/just part of the explanation for, say, women's concentration in nursing, for workers of both sexes in the sex trade? Is the researcher bound to agree with the worldview of the people with whom she/he works so closely? Can an 'outsider', as Waquant was in the world of Black boxers, produce as satisfying an ethnography as an 'insider', as Desmond insisted he was? These are all important questions. They are particularly relevant in this chapter, but are also part of a critical assessment of the theoretical significance of ethnographic and case studies of different forms of work. While empirical generalization is clearly not the aim of workplace ethnographies, they do make a significant contribution to the development of theoretical understandings of the social construction of identities at work (Edwards and Belanger 2008), as well as providing insights into often unfamiliar worlds of employment. I return to some of these questions in chapter 9.

Part III
High-Touch Servicing Work in Specialist Spaces

Plate 4 Institutionalized care work: the commodification of women's 'natural' skills. Photo © Don Smetzer/Getty Images.

7

Bodies in Sickness and in Health: Care Work and Beauty Work

'Ah', sighed Mrs Gamp ... 'what a blessed thing it is to make sick people happy in their beds, and never mind one's self as long as one can do a service.'

Charles Dickens, *The Life and Times of Martin Chuzzlewit*, 1844

In our culture, not one part of a woman's body is left untouched. No feature or extremity is spared the art or pain of improvement.

Andrea Dworkin, *Woman-Hating*, 1974

One of the main features of bodies is their contradictory tendency. On the one hand, bodies age, decay, wither and ail, but on the other hand they are also open to improvement through exercise, diet and forms of physical manipulation, actions that may delay the effects of ageing. The search for improvement is often driven by an idealized notion of a body beautiful: a notion that is historically and geographically specific, as well as gender specific and sometimes varying by social class. In both sets of processes, close and personal interactive body work, by specialists and by the individual 'owner' of the body in question, is a key part of daily and generational bodily maintenance. A range of jobs and professions is involved in the care for sick and/or elderly bodies and in the maintenance and improvement of healthy bodies, many of which are constructed as particularly appropriate occupations for women, associated with those typically female attributes of love, care and empathy that are widely seen as natural feminine attributes. As I argued earlier in the discussions of domestic and sex work, feminine skills in relational work are too often taken for granted. In this chapter, as I shall show, nursing (at least until recently), care work and work in salons and beauty parlours are all feminized occupations that are correspondingly poorly paid. But not all body maintenance work is feminized or low paid. The distinction between doctors and nurses is a key example of the lack of association between body work and femininity (Pringle 1998). Through the social construction of medicine as a technical and scientific occupation, it is disassociated from direct work on the bodies of others and, with its links to elite male workers, is constructed as a high-status, well-paid occupation. Other types of body work lower in the social-status hierarchy – personal trainers and sports coaches, for example – are as likely, if not more so, to be undertaken by men rather than women. The association of masculinity with strength, power and fitness enables men to construct these types of work as congruent with their sense of themselves as masculine.

In this chapter I first address the connections between embodiment, high-touch body work, femininity and social status in caring occupations through the examples of nursing and care work, exploring the associations with femininity and the paradoxes that result for men in these occupations. Here, the bodies of the clients/customers/patients typically are unwell and/or ageing and the relations of power often, although not always, privilege the care providers. I then turn to the healthy body, looking at professions that are concerned with bodily improvement and with adornment, beautification and display: traits that are often associated with female vanity.

Bodies, Emotions, Care and Femininity

In chapter 4 the ways in which stereotypical attributes of femininity are mapped onto occupations, constructing the skills developed in undertaking domestic tasks and childcare as 'natural' attributes of women and so not 'real skills', were outlined. The nature of this work and the limited potential to increase its productivity was explored as an additional reason for the typically low levels of financial recompense for workers employed in the domestic sphere. In caring work in the public arena – in hospitals, hospices and elder care homes – the same sets of associations and the need for a high staff-patient/client ratio to establish and maintain interpersonal relationships are also a significant part of the social construction of nursing and care work as feminized, although there is a significant status differentiation between them. Nursing is what was once termed a semi-profession (Etzioni 1969): the condescension in the label 'semi' associated with the dominance of women in occupations so categorized (Hearn 1982). Librarianship and social work, for example, are also included under this rubric. Clearly, these are demanding professions, based on sets of specialist skills and are usually graduate-only entry, but their reliance on a predominantly female labour force reduces their status. Nursing has also moved towards graduate entry in the last decades and its status has increased to some extent, although it is still poorly paid. Care work remains a 'high-touch', low-status job, including within its ranks large numbers of unskilled workers without relevant credentials who undertake the 'dirty work' end of bodily care. Increasingly, however, migrant workers with nursing qualifications take jobs within care homes, as entry into health and health-related professional jobs is often closed to them.

The Social Construction of Nursing: The Lady with the Lamp

One of the paradoxes of nursing is the way in which, as it moved from the home to the hospital, what is often a hard, dirty job involving heavy lifting, bodily fluids, smells and intimate care became constructed as a job suitable for middle-class women. In part, this is because of the associations between care, love and emotions which permit the care of the sick to be constructed as an act of charity and compassion, as much as an occupation or profession. Since the time of Florence Nightingale whose work in the Crimea has entered popular myths about nursing, the concept of caring has been central to the construction and representation of nursing. Care for others is associated with women's 'natural' talents of empathy and sympathy, able and willing to perform tasks of social reproduction and bodily care that are often associated not only with the body but with the private, the personal and the intimate realms of embodiment where, of course, the relationships between the cared-for and the caregiver are based on love and the ties of affection, rather than on a monetary exchange (Novarra 1980; England and Folbre 1999).

When caring is provided in the public arena – whether by the state on the basis of a system of rationing and entitlement or in the market on the basis of ability to pay – the association between caring and affection leads to a paradox. Commodified caring labour is both highly valorized (as more than a job, as a vocation) and devalued and correspondingly poorly remunerated. As Julia Twigg (2000a) noted, there is a direct parallel between this paradox in commodified caring labour and the social construction of women's unpaid caring work in the home. In each case, the social roles and labouring that are involved are at the same time both 'sidelined, lowly regarded and unsupported' and represented as being of 'supreme value' (p. 407). Twigg identifies what she terms a 'halo effect' (from Nightingale's lamp perhaps?) which obscures the hard physical labour undertaken, illuminating only the tender provision of care, and so skewing the debate about the nature of the work involved. Workers are often represented as having a 'special vocation' involving innate skills and quasi-religious devotion (think of the parallels between old-fashioned nursing uniforms and those of women in religious orders), contributing to the low status and poor pay of this work (England 2005).

In an interesting case study of the caring labour involved in looking after dying patients in a hospice, Nicky James (1992) argued that caring in the public arena comprises three components:

- organization
- physical work
- emotional work

The organization – in the sense both of the institutional setting and the ways in which tasks are organized – sets the framework within which care is carried out, perhaps in a quasi-familial setting in a hospice, but more typically in a bureaucratic setting in a hospital. Nursing, like the domestic work, combines physical and emotional labour. Numerous case studies of caring occupations, including nursing and elder care, have shown that carers themselves adopt this binary distinction when talking about their work. Nurses tend to refer to the different tasks that they perform as either 'work' or 'care': the latter involves emotions and feelings of attachment to those for whom they care; the former encompasses physical labour. Thus, as James (1992: 497) suggests, 'the framework of physical labour … became the justification and explanation of paid work. Having been sitting talking to a patient a nurse would say "I must go and do some work now," meaning physical tasks.' This distinction was used by staff to construct boundaries, allowing them to complain or to take industrial action, to withdraw their labour without feeling they were letting their patients down, but it was also part of the maintenance of the assumptions which devalue their labour through its associations with 'love' and affection.

As I argued in chapter 4, the boundary between physical and emotional work is not always easy to draw as the tasks overlap. A range of emotions are involved in doing the dirty work, including the management of their own emotions by carers and nurses – the distaste or disgust that certain tasks, material or odours arouse. A typical way of coping with these emotions is the development of distancing mechanisms to separate workers from their work. This displacement mirrors the tactics used by sex workers discussed in chapter 5. Before exploring these distancing mechanisms in a case study of domiciliary care workers, I explore the organization of care work and the stress involved in performing emotional labour in this example of embodied interactive employment.

Empathy and Stress in Customer-Oriented Bureaucracies

Doing emotional labour is an important part of all sorts of interactive service work (Hochschild 1983; Krumal and Geddes 2000; Williams 2003; Hampson and Junor 2005). Face-to-face encounters between workers and consumers/ clients require workers to produce a 'display, deployment, and management of emotions' (Hughes 2005: 604) as part of their day-to-day work. Like other forms of consumer service employment, nursing involves a three-way interrelationship – with superiors, with colleagues and with patients (Wolkowitz 2002). Hospitals are a good example of a 'customer orientated bureaucracy': a term that captures the ways in which workers are governed not only through the standardizing disciplining of a bureaucratic organization, but also by the demands of customers (patients) who are physically present (Korczynski 2001; Kerfoot and Korczynski 2005). The embodied attributes and gendered performances of workers are particularly important in customer-orientated bureaucracies, as workers become responsible for resolving the intrinsic tensions between the rational organization of their work and the more unpredictable demands of patients. 'Empathetic' women workers have a particular role in easing customers (and patients) through the standardized service exchange (Forseth 2005). In the next chapter, I look at the ways in which the same sort of relationship between gender, tension and empathy works in relationships with guests in hotels; as the previous chapter illustrated, albeit in a different type of setting, door staff also have to resolve tensions between customers and between customers and employees. In the healthcare sector, women workers are essential in the resolution of some of the tensions inherent in circumstances where the 'customers' are anxious, sick, perhaps elderly and sometimes confused (Halford 2003; Bolton 2005). Interactions between the patients, their relatives and the workers take place in the context of gendered understandings of care in which nurses are considered to provide care and empathy, as well as bodily maintenance, whereas doctors provide scientific or rational information and so are not relied on for emotional support in the same way as nurses are.

As I have already argued, the emotional demands of this part of caring labour tend to be unrecognized as a skill, and so financially poorly rewarded (Smith 1992; Payne 2006), despite the evident importance of emotional labour in the satisfactory delivery of care. This is not a new claim, however, nor is nursing and careworking a new form of work. Its significance has grown, however, in an economy where women are increasingly involved in waged labour and so less available to care for their dependents outside institutional provision and because of the needs for care of an ageing population. Half a century ago, Isobel Menzies Lyth (1959), a psychoanalyst, wrote a controversial paper that was largely ignored until it was republished in 1988, in which she prefigured recent arguments about emotional labour, as well as the significance of hospitals' organizational practices. Based on research in an English teaching hospital in the 1950s, she showed how the organization of nursing and the intimate relationship it demanded with patients impacts on the organization of care, leaving those closest to patients – junior nurses and nursing assistants – exposed to emotional pressures that more senior staff and managers were protected from. She discovered extremely high levels of distress and anxiety among junior nurses. As Menzies Lyth argued, this stress is a product of the fact that the work of nursing is itself exceptionally anxiety producing. Nurses work with people who are ill or dying in circumstances where incorrect actions might have devastating consequences. Nurses must respond both to the distress of patients themselves and to the distressed family of the patient. Furthermore, many of the tasks involved in nursing the sick are distasteful or repulsive and so place emotional demands on the workers carrying them out.

Menzies Lyth documented the ways work in hospitals in the 1950s was organized to contain and modify nurses' anxiety. There was a dominant belief that if the relationships between nurse and patient were (too) close, the nurse would experience more distress when the patient was discharged or died and so work practices were developed that would encourage distance. Nurses were required to perform a few specialized tasks with a large number of people, rather than providing continuous care for a smaller number of patients, in a manner more reminiscent of a Fordist form of organization than the niche production of specialist care for the 'whole' body. Calling patients by their condition – the tumour in bed 19 for example – rather than by their given name was also common. Similarly, the weight of responsibility for making a final decision was mitigated in a number of ways. Even inconsequential decisions were checked and rechecked and simple tasks were passed up the hierarchy of command. There were no mechanisms in place to support nurses who became distressed. The administration did not, for example, acknowledge that a patient's death affected nurses or provide support to deal with this and other forms of emotional distress. Instead, the rationale developed that a 'good nurse' was 'detached'. There is an interesting echo here of the advice that used to be given to social researchers.

Menzies Lyth found, however, that although these distancing mechanisms protected nurses from some anxieties, they also created new ones. In interviews, nurses expressed guilt about many of the bureaucratic procedures. Even though they might have carried out their instructions exactly, they still believed that they had practised bad nursing, arguing that they were caring for the system's rather than for patients' needs. Menzies Lyth's arguments pre-date by many years the insistence by James and others (Kang 2003; Seymour and Sandiford 2005) on the importance of the context under which caring occurs, but remain absolutely relevant. The same sorts of distancing mechanisms are evident in, for example, the long training and professional ethics central to medicine and specialist nursing which provide ways for middle- and high-end workers to deal more easily with their own emotions and the emotional interactions with patients. Other mechanisms such as exposing only part of the body and hiding the rest of the patient under a sheet or a screen are also used to reduce the intimacy of interactions. Such resources and mechanisms are not generally available to more junior nurses, to healthcare assistants or, especially, to the unskilled carers in institutions such as care homes for the elderly, where the careful manipulation of emotions is an essential way of reducing the degradation of certain intimate procedures.

It has been recognized that auxiliaries, healthcare assistants and even hospital cleaners, whose work would seem to fall centrally on to the physical labour side of James' distinction, also undertake a great deal of emotional labour during the course of their daily tasks (James 1992; Sass 2000). This might involve taking time to talk to someone as they bathe them, serve their meals or clean their room. These jobs, although they are defined as unskilled and menial, often require skill and the exercise of judgement and, as Glenn (1992) noted, involve the transfer of domestic or maternal skills into the sphere of waged work. Turner and Stets (2006) have argued that the demands of emotional labour may be particularly stressful for recent migrants, who are a key part of the workforce in caring occupations in many British towns and cities (May et al. 2007). Non-British born migrants working in low-status and poorly paid jobs in the service sector often find that they must perform both emotional labour and physical care across a cultural divide, undertaking tasks that they do not regard as appropriate or acceptable parts of the job, or providing emotional support for, say, elderly people, more usually undertaken by family members in their countries of origin (Dyer, McDowell and Batnizky 2008).

Caring as Body Work/Dirty Work

The provision of care is also hard physical labour that often involves close and intimate work on the bodies of others. The human body, particularly when ill, old or diseased, unsettles the modern western emphasis on an individual's rational autonomy and the supposedly bounded nature of the body. Although there are exceptions, such as psychoanalysis and counselling, most caring work takes the customer's body as its immediate site of labour. In a culture which esteems the cerebral over the physical, the autonomous above the dependent, and the disciplined over the uncontrolled/able, 'body work' in the caring professions is marked by the intimate and often messy contacts that are part of the interactions between labourer and embodied client (Gubrium 1975). Work is not only performed on the bodies of others, but it is often undertaken using the bodies of the workers as a primary vehicle (Jervis 2001). Patients have to be lifted and moved, turned, washed, walked and massaged, typically by the lowest-status workers. And as nurses are promoted 'they move away from the basic bodywork of bedpans and sponge baths towards high-tech, skilled interventions; progressing from dirty work on bodies to clean work on machines' (Twigg 2000a: 390), neatly encapsulating Brush's distinction between high-touch and high-tech work discussed in chapter 2. It is the lower-status nurses and nursing assistants who are left 'to deal with what is rejected, left over, spills out and pollutes' (Wolkowitz 2002: 501), often hiding this dirty work away behind screens (Lawler 1991). For the lowest-status care workers, body work often involves caring for elderly and/or diseased bodies, regarded as loathsome or feared bodies in contemporary western societies that valorize young, slim and unmarked bodies (Young 1990; Lawler 1997; Jervis 2001). Not for nothing is geriatric care termed the Cinderella service in the British health service.

Undertaking the 'dirty work' of interactive body work represents a challenge to workers beyond their low pay, as the symbolic associations that contribute to the structuring of body work as being of low worth become attached to the workers themselves. Thus, those involved in this type of work may become identified as 'dirty workers', as their waged work poses a challenge to their sense of identity and self-worth (Bates 2005; McGregor 2007), although workers are able to adopt a number of strategies to attempt to mitigate their labelling and the distasteful or intimate parts of their jobs, through, for example, the use of humour, by forming strong collective identities and by emphasizing the dignity of earning a wage (Bolton 2005; Stacey 2005; Kreiner, Ashforth and Sluss 2006). Body work is, perhaps above all, 'ambivalent work' (Twigg 2000b: 391), as the distinction between emotional and physical work indicates. It involves pleasure and emotional intimacy, but also evokes disgust, the necessity of dealing with human waste, or touches on taboo areas around sexuality and ageing. The extent to which emotional work brings stress or satisfaction has been the subject of a long debate (Hochschild 1983; Steinberg and Figart 1999; Erickson and Ritter 2001; Williams 2003; Bolton and Houlihan 2005; Forseth 2005; Zapf and Holz 2006).

In the next sections, I draw first on Julia Twigg's (2000a, 2000b) study of peripatetic care workers to show how they manage this ambivalence and what sources of satisfaction are found in this work. I then turn to the position of men in nursing, looking at the ways in which masculinity may advantage men in the profession, despite the associations between caring and femininity, ensuring their rapid movement from the dirty parts of body work into management. I end this part of the chapter with a discussion of Tim Diamond's (1992) study of a residential care home in Chicago to explore wider questions about the relationships between class, gender and race in the construction and segmentation of a caring labour force.

Providing Care in the Community

Like nursing, care work undertaken in the community – that is, outside the specialized settings of hospitals or hospices – is gendered work, typically, although not always, undertaken by women. It is also low-status, hard, heavy physical work that, nevertheless, also involves the emotions. In her book about the employment world of community care workers, Julia Twigg (2000b) opens one chapter with a quote from Bates (1993: 14) that so nicely captures the different aspects of care work that I have reproduced it here:

> To the outsider, the 'caring' world may have appeared as one of pink overalls, cleanliness, jangling keys, 'come along Annie', wheelchairs, walks around the grounds, trolleys, laundry, the ubiquitous institutional smell, dozing residents in television lounges. This was the re-presented, more acceptable face of the occupation, which in effect in terms of contemporary cultural constructs, was a 'heavy', 'dirty' job, steeped in taboo subject matter, such as the body, age, 'shit shovelling' and death. These social responsibilities were transported form the wider society to a sub-society, staffed largely by women. Within this sub-society, a partial and truncated version of 'caring' was enacted, involving constant tension between caring for and processing people.

This description is of a community home for elderly people, but commodified 'community' care is also provided in people's own homes. Here care work involves not only caring for the bodies of clients, who usually are older or disabled people, in the holistic sense of making them comfortable, providing some human contact and sympathy, but also managing the tasks involved in the daily maintenance of clients and their bodies. Thus, it is a type of work that involves transgressing the boundaries of normal social life, especially when the work takes place in clients' private homes. In this sense, this case study also fits into the material discussed in chapter 4. I have included it here, however, as the workers are peripatetic; they are not restricted to a single home as nannies are and because their work is more akin to nursing in that they care for declining or sickening bodies rather than for children. One of the most complicated challenges of this type of care work is how to manage the relationship between the emotional aspects of caring and the disgust that is associated with some of the essential parts of the job, especially when they are carried out in spaces that are not specially equipped to facilitate the efficient provision of the care and maintenance of human bodies. Care in the home thus touches on, even encroaches on, some of the most significant issues in people's lives – their feelings about their body, the meaning of home (Blunt and Dowling 2006) and the maintenance of dignity in ageing and dying (Lawton 2000).

Doing care work in the community: location, class and ethnicity

Who takes a job as a care worker in part depends on the structure of the local labour market and the range of other opportunities for people – mainly women – with few skills or educational credentials. It is also related to the nature and structure of the migrant population, both within local labour markets and in the UK as a whole. Care workers typically are middle-aged,

Table 7.1   Non-UK born care workers 2006

	UK	*Non-UK*	*% Non-UK*
Care assistants and home carers	535,000	105,000	16
House-parents and residential wardens	35,000	2,000	5
Nursery nurses	151,000	7,000	5
Nursing auxiliaries and assistants	197,000	23,000	10

Source: Office of National Statistics 2006

working-class women, although their numbers are often swollen, especially, although not only, in London, by migrant workers. These include both people of colour and a group of more casual and transient workers: young, often white, women from countries such as Australia, New Zealand and South Africa who fund their travels in Europe by doing care work on a temporary basis. Since 2004, growing numbers of A8 migrants have also taken jobs as domiciliary or residential care workers. Care work is also racialized – in London and Birmingham, for example, a significant proportion of the labour force is Black; in other cities older white women from Ireland are also in the labour force, reflecting the history of migration into the UK in the postwar period. Table 7.1 shows numbers and the proportion of workers undertaking different aspects of care who are non-UK born.

The highest percentage of the non-UK born social care workforce (which does not include state registered nurses) is from Zimbabwe (12 per cent), followed by the Philippines (10 per cent), Ghana (7 per cent), Poland (6 per cent), Germany (6 per cent), Nigeria (6 per cent), India (5 per cent), Jamaica (3 per cent) and the Irish republic (3 per cent). Most of these workers have joined the labour force in the last ten years. The impact of the new accession countries is visible in data from the Workers Registration Scheme (WRS) with which newcomers from the accession states are obliged to register when they take employment. Between accession in May 2004 and June 2006, 12,610 people registered as care assistants and home carers, of whom the majority (62 per cent) were from Poland. In London, where non-UK born workers are a third of the total labour force, their share of care work is significantly higher – 68 per cent of all care workers in London are foreign-born. Care work is low-status work and often the last option for those with the most limited choices in the labour market. As Bates (1993) and Skeggs (1997) have argued in their work on young school leavers, caring courses are the least well regarded of all training options open to young women with few qualifications. Accepting a place is usually a response to poor/low labour market options rather than a positive decision. The young women on these courses tend to keep quiet about the nature of their work, before and after training, and, as Bartoldus, Gillery and Stuges (1989: 207) noted in a US study of caring, workers believe the general public sees them 'as "unskilled maids" who did society's "dirty work"'.

What is unusual about domiciliary care work – the example that Twigg analyses – is that it is almost entirely unsupervised on a daily basis. This is atypical among low-skilled interactive jobs in the health service, although it is common in some of the other forms of 'home-based' work discussed earlier. Domiciliary care workers have a greater degree of independence than, for example, auxiliary workers and cleaners in hospitals and perhaps as a consequence greater prospects of job satisfaction. They are often able to arrange their hours around the time demands of their own daily lives, fitting in shopping or picking up children from school, for example, between visits. The work involves a variety of tasks, not all of them unpleasant (as the quote from Bates indicates), but it also often involves fragmented tasks and visits and so may be difficult to organize so that time is used most efficiently. Split shifts are common, for example, and so work expands across or beyond a conventional working day. And as I noted earlier, because the work takes place in people's own homes, it is often carried out in cramped or unsuitable conditions and spaces: in people's private bathrooms, for example, where lifting aids and other equipment often do not fit into the space available. Like the basic tasks of nursing, domiciliary care work thus also places demands on the workers' own bodies as heavy lifting or bodily manipulation in awkward spaces create stresses and strains.

The dirty work of keeping clean

One of the main parts of domiciliary care work involves keeping the bodies of clients clean. This is the focus of Twigg's (2000a, 2000b) analysis of the dual aspect of care work as dirty work and emotional labour. Her study was based on interviews with 30 older and disabled people, 24 care workers and 11 front-line managers, living and working in an affluent area of inner London and in a more deprived coastal town. The clients included men and the workers were mainly women: 31 compared to just 3 men, a ratio that reinforces the associations between femininity and care. All the participants were interviewed – some alone, others in small groups (Twigg 2000a: 394–5) – but none of the interviews took place during the interaction between the carer and the cared-for. Clearly, in this type of research, issues about privacy preclude a close and intimate engagement of researchers in the labour process, unless, of course, the researcher is also a care worker. In the second case study of bodily care for elderly people that I discuss below, this was the case, as Tim Diamond (1992) took a job in a nursing home in Chicago. His study raises different questions about the research process than does Twigg's work, but together they provide a moving insight into the labour and love involved in caring.

In her study, Twigg explored in detail two key aspects of the bodily interactions involved in care work: first, dealing with dirt and emotions of disgust, and secondly, dealing with the naked bodies of others. Care work, as Twigg (2000a: 389) argues, 'is about dealing with human wastes: shit, pee, vomit, sputum': all products of the body that usually are evacuated in private, although there are cultural differences between societies. Spitting in the street, for example, is commonplace and acceptable in some Chinese and Portuguese cities, but is frowned on elsewhere. In all societies there are social systems and accepted norms for dealing with waste. Disgust enters the equation only when these are breached. As Mary Douglas (1966) noted more than forty years ago, 'dirt is matter out of place'. In a clean bathroom where waste is neatly deposited and flushed away, its 'dirtiness' is not an issue. For elderly incontinent clients, unable to reach the WC, it is, and dirtiness is associated with shame.

Dealing with excrement was the part of the job that evoked the greatest sense of disgust, in part because of the smell:

> Someone in a mucky bed, you know, you just – you can put all the protective clothing on, gloves, your plastic apron and everything, but you sometimes feel unclean. (Susan in Twigg 2000a: 396)

> The smell just stays with you. (Vicky in Twigg 2000a: 396)

Smell, Twigg argues, is so significant as 'it extends the patient's corporality in such a way that intrudes and seeps into others' spaces. Odours by their very nature cannot be easily contained; they escape and cross boundaries' (p. 397).

Workers also disliked other aspects of bodily maintenance work: some disliked dirty teeth, others flaking skin, people being sick, or the sound of expectoration:

> What gets to me is when they're coughing up phlegm and put it in a bowl. (Tracey in Twigg 2000a: 394)

As Lawton (1998) noted in her work in a hospice, these patients' bodies have become unbounded by virtue of their inability to control their waste products. This challenges western notions of the singular, separate, inalienable body of each individual. In hospitals, hospices and in the home, carers struggle to combat dirt and odours in an attempt to reassert dominant 'ideas about living, personhood and the hygienic, sanitized, somatically bounded body' (Lawler 1991: 123). As I argued in the introduction, this idealized separate body is closer to a masculinized version of the body. Women's bodies have long been represented as unbounded, messy, leaky and fecund (Grosz 1994). Menstruation, pregnancy and lactation all challenge this idealized singular, bounded and separate body. Iris Young's (2005) discussion of her own pregnancy includes an example of the unbounded body. She noted that academic colleagues felt free to touch her and the 'bump' in ways they would not have dreamed of when she was not pregnant. And, as I also noted earlier, the efforts to deodorize and/or re-perfume the female body is the basis of a huge industry, selling a vast variety of deodorants and fragrances not only to women but increasingly to men. In the second part of this chapter, I explore the work of beauty therapists who play a key role in the beautification of bodies.

The second arena of visceral emotion and disgust explored by Twigg concerns the feelings generated by the sight and the washing of elderly bodies. The hegemonic idealized body that dominates not only advertising but the labour market is a slim, fit, young – and usually white – body (Bordo 1993). Western culture provides few – and even fewer positive – images of ageing bodies and almost never of ageing naked bodies. Twigg reports that her respondents found the initial sight of older people without clothes surprising and often shocking:

> I just had to stop myself staring at people, because I hadn't really seen ... because you don't really see people naked. (Sophie in Twigg 2000a: 397)

Sophie was also surprised at the variation between older people's bodies. Other workers emphasized the differences between their own, still young bodies, which led them to reflect on the inevitable loss of strength and vitality associated with ageing, and, more specifically, on the implications of the loss of sexual attractiveness. The nakedness of older people thus 'offered a glimpse into a personal future that was discouraging and unwelcome' (Twigg 2000a: 398).

One of the most challenging aspects of domiciliary care work was the need to touch older people's bodies. Intimate care work inevitably involves skin-to-skin contacts, although many care workers wore gloves for at least some of the tasks. Here is Sophie reflecting on the touch of older people's skins:

> It was really weird when you feel the skin ... actually to feel someone's skin it's quite different than having the glove over your hand. It's *much* more intimate. (Twigg 2000a: 294; original emphasis)

The use of gloves while bathing clients was a common way of avoiding too much skin contact and intimacy. They acted to construct 'a barrier of professionalism between the client and the worker' (p. 404). Sophie, also suggested that the use of gloves was a way to avoid the associations between skin contact and sexual pleasure:

> I appreciate wearing them [gloves] ... you know they [male clients] might get into it having a nice girl rubbing their back with bare hands ... having the gloves,

> it's kind of like wearing a badge saying: I am here for a reason, this is what I am doing. I am not just here with you naked, washing you. (Twigg 2000a: 404)

In some cases, however, the wearing of gloves was constructed by clients as somehow demeaning or dehumanizing, 'evoking a sense that the person being handled is contaminated or subhuman' (p. 404) and the association with how police and other officials handle people who are suspected of being HIV-positive increases this sense of being demeaned.

Dealing with both dirt and naked intimacy are more often associated with relational ties of affection – between mothers and children, for example, or between lovers – than being part of a commodified exchange, and so both carers and their clients need to find ways to cover or get beyond embarrassment and to distance themselves from the intimate body working that is taking place. A number of strategies were adopted by the workers whom Twigg interviewed. Some care workers attempted to put themselves in the position of the recipient of care and to use empathy to overcome their embarrassment. Avoiding direct language was common, as well as chatting or joking while dealing with difficult parts of the job. Some of the women care workers dismissed the notion that such work was necessarily embarrassing, seeing dealing with dirty bodies as a natural extension of women's lives as mothers. Twigg also noted that most care workers did not wear uniforms and so presented a somewhat ordinary, even homely, appearance in contrast to nurses: a fact that was valued by many clients.

The type of body work carried out by these care workers is low-paid, low-status and, to a large extent, demeaning work for both the worker and the client. As I suggested earlier, the associations of dirt are often assumed to transfer to the bodies of the workers themselves. In response, workers emphasize the more pleasant 'emotional labour' that is a large part of their jobs. The care workers interviewed by Twigg, for example, made a distinction between emotional closeness and intimate touching work. They told her that they often developed an emotional attachment to their clients and were warm and tactile in interaction, but they emphasized 'the social touch of cuddles and kisses, rather than the intimacy of bathing, as a route to emotional closeness' (Twigg 2000a: 399). And it was the emotional closeness rather than the 'dirty work' that workers emphasized in discussions about their working lives with friends and relatives. Developing close ties and providing comfort to clients was seen as one of the advantages of the work. As Barry, one of the three male care workers whom Twigg interviewed, told her:

> I get enjoyment, I have to admit, that I get enjoyment out of it and pleasure … seeing the pleasure on somebody's face when you walk through the door first thing in the morning…. The pleasure of seeing the client's face and knowing that, when you leave, they're happy. (Twigg 2000b: 164–5)

Although Twigg does not comment, I wonder whether the sense of reluctance to admit to pleasure in caring in Barry's comment is a reflection of the more typical association with these tasks with women and femininity. As Twigg (2000a: 408) notes, women's association with body work derives from 'the greater freedom that women are accorded in their access to bodies compared with men', at least if the men are not professionally qualified as doctors. Hegemonic notions of sexuality construct men as predatory and women as passive, so necessitating greater restrictions on men's access to bodies. Typically in the UK, domiciliary male care workers are restricted to dealing with men's bodies, whereas women look after both sexes.

The emphasis on the satisfactions of emotional labour and the felt need to hide the dirty body work involved in caring labour means that this latter part of the job is undervalued. Further, the establishment of emotional ties with clients, while bringing job satisfaction, is also a problematic aspect of care work. Becoming (too) attached to clients is an occupational hazard and some care workers interviewed by Twigg found it hard to bear the unhappiness of their clients and their growing dependence on them. They also had to learn to deal with death, often unexpected and in the home, rather than in a specialized or sequestered location, such as a hospice. Workers talked about having to learn to switch their emotions off when they leave work and to accept that care work, as a job, cannot be unbounded and unlimited. As Twigg (2000b: 170) noted, 'their jobs and the incentives in relation to them, are structured by economic rationality.... The work thus does have clear limits in the sense that care workers are only allocated and paid for specific periods of time and often only for specific tasks.' All care workers have to negotiate the tension between 'the ethic of care' based on love and ties of affection and the economic rationality of waged labour in order to avoid becoming the 'prisoner of love': a dilemma identified by Folbre and Nelson (2000).

In the next section I explore the ways in which male nurses are able to negotiate this tension and construct their work as appropriately masculine when working in a profession saturated with images of femininity. I then return to care work, although in an institution, to explore how the associations between care and femininity construct male workers as out of place in a nursing home in Chicago.

What About Male Nurses?

As I have argued earlier in this chapter, caring is constructed as an essential female attribute and so an ideal of feminine caring is mapped onto nursing and care work as a natural talent embodied by women rather than a skill constructed through training. Indeed, a key theme of the book, as is now abundantly clear, is the way in which the attributes of masculinity and femininity are mapped onto jobs and occupations to include one gender and exclude the other. As we have seen, women are 'naturally fitted' for certain types of work and not for others, while women who find themselves in masculinized occupations often suffer from being the 'token' woman, marginalized, subject to exclusionary or hostile behaviour and sometimes sexual harassment, and so suffer career disadvantages (Kanter 1977; Collinson and Collinson 1996; McDowell 1997; Simpson 1997). But what of men who enter female-dominated occupations? Do they suffer from being a 'token' man? Are they similarly marginalized and harassed or is their visibility in the workplace an advantage? Although there is not an enormous literature about men in non-traditional occupations, there does seem to be evidence that they benefit from their position through assumptions about career development and leadership potential associated with stereotypical views about men's attitudes to and aptitudes for waged work. Men in primary school teaching, for example, may find they have advantages in the recruitment and promotion process and they often ascend the career ladder more quickly than their women colleagues, but their advantage may alienate them from female staff (Williams 1993). Male nurses, too, are often promoted more rapidly than women (Bradley 1993).

In a profession such as nursing which is so heavily imbued with notions about emotional labour, men might be expected to be at a disadvantage as the oppositional construction of rationality and emotions which maps onto a male/female divide seems to construct them as out of place. Men – currently 10 per cent of nurses and 14 per cent of primary school teachers in the UK – may, for example, find their competence challenged if they rely on versions of traditional masculinity at work or alternatively find their identity and/or their sexuality challenged if they adopt what are constructed as feminine attitudes and behaviours at work, whether in the classroom or the hospital ward (Heikes 1992; Isaacs and Poole 1996; Evans 1997; Alvesson 1998). Lupton (2009) has argued that men in female-dominated jobs and workplaces fear feminization and stigmatization, not necessarily or only in the workplace but also among their peers and in their social networks. As Kimmel (1994) has argued, men scrutinize each other for signs of femininity and homosexuality. For most men, constructing a satisfactory hegemonic masculinity demands continual self-scrutiny and hard work. Manhood is always insecure, uncertain and incomplete (Connell 2000) and in feminized occupations men's anxiety about their masculine identity is exacerbated.

In an interesting study based on 40 in-depth interviews with men working as primary school teachers, cabin crew (one of Arlie Hochschild's (1983) examples in her classic work on emotional labour), librarians and nurses, Ruth Simpson (2004) found that men in these jobs adopted a range of strategies to maintain and reproduce behaviours that were congruent with their sense of themselves as masculine. These included re-labelling, status enhancement and distancing themselves from the feminine. I draw on her interviews with 15 nurses in 10 different hospitals in the southeast of England, who responded to an advertisement the author placed in a professional nursing journal. Five of the men worked in general nursing, five in mental health, four in accident and emergency and one in palliative care, and the site and type of work also affected social relations and the construction of appropriately masculine identities in the workplace. Masculine authority and so an absence of emotions was easier to assert and more of an advantage in the Accident and Emergency department than in the other arenas of the hospital. Male nurses there emphasized that the work was exciting, fast, stressful and hard to handle compared to general nursing, whereas male mental health nurses emphasized the links with the custodialism of the job, which was represented as a symbol of masculine authority (Simpson 2004: 360).

The nurses whom Simpson interviewed recognized that their minority status gave them advantages. Four key themes emerged from the interviews that she labelled:

- the career effect
- the assumed authority effect
- the special consideration effect
- the zone of comfort effect

Many of the nurses believed that they had greater opportunities to move up the hierarchy and eventually into management (where as I noted earlier the ability to perform emotional labour becomes less significant). This career effect often operated through the special consideration effect when men were given opportunities to acquire skills and expertise that were not open to women nurses. Simpson (2004: 356) included a comment from one man that during his training:

> I always got first crack at 'does anyone want to go to the theatre to see this?', 'would someone like to accompany the doctors to do that?'

But others among her respondents argued that the assumptions that men would prefer to move rapidly into management were not necessarily helpful. This man was uncomfortable with assumptions by his (female) colleagues that he wanted quick promotion. He reports staff nurses saying to him:

> 'Oh, you won't stay at D [the lowest grade] for long, they never do, the men.' And I am thinking what's wrong with just being a normal nurse? Why should a man have to go up to management level? (Simpson 2004: 356)

The men who talked to Simpson also told her that they seemed subject to different and more relaxed rules and expectations, often able to get away with mistakes or behaviour that would have earned a woman a reprimand. The assumed authority effect was also both an advantage and disadvantage. Men were assumed to be more senior but also more assertive and sometimes expected to speak out on behalf of their colleagues and patients in ways that women were not, or take on tasks that may be outside their expertise. Here's a newly qualified general nurse talking about this:

> They all seem to come to me rather than Tracy [a more experienced female colleague] and even Tracy will come to me and ask for advice when she is more than capable of knowing herself. But sometimes you are frightened that you're going to say the wrong thing all the time because I don't have the knowledge base. (Simpson 2004: 358)

Few of the men Simpson talked to reported feeling isolated or uncomfortable at work, even though they were greatly outnumbered by their female colleagues. Indeed, most of the male nurses reported feeling relaxed and at ease working with women. They appreciated their support and valued their assimilation into the 'realm of women' (p. 358). Simpson captures this in her concept of the 'comfort zone'. However, when men were asked to reflect on the popular images associated with their job, on how it fitted with their own identity and with dominant notions of masculinity, many of them reported strategies to try to overcome the sometimes evident incongruity between their own self-image and the feminized image of their job, including re-labelling the job in discussions with friends, recasting its main characteristics to emphasize more masculine components and distancing themselves from 'the female' and from women in the workplace. As Leidner (1993) found in her study of insurance agents, similar processes of emphasis allowed men to construct their work as congruent with their sense of masculinity. Thus a nurse who had worked with drug addicts, for example, emphasized the danger of the work that took place in 'a sort of rough and ready area ... around quite tough issues', as well as his own masculine appearance: 'I'm tall and big and unshaven and got a big scar on my face' (p. 359).

Several nurses reported anxiety generated in discussions with friends and acquaintances about their work:

> Men ... they laugh and say 'oh, you're a man for goodness sake, why have you chosen to do nursing? Why can't you be a doctor?' (Simpson 2004: 361)

But not all men want to enter traditionally masculine-coded occupations, nor as Simpson concludes did all the men that she interviewed working in female-dominated occupations want to adopt conventional masculine attributes and career aspirations. What emerged from her work are the sets of contradictions and paradoxes that faced these men as well as the variations between them. While some of them valued the opportunities to get ahead, others preferred to resist the associations between masculinity, authority and ambition. Nevertheless, it seems clear that for men there is greater freedom and more opportunities in non-traditional occupations than there is for women in a similar position in male-dominated jobs. As Simpson (2004: 363) notes, 'the positive cultural evaluation given to male attributes in society [is] reflected in the rewards that accordingly accrue' to men in non-traditional work.

Gender, Ethnicity and Race in a Care Home

A more personal and in-depth study of the position of a man in a female-dominated occupation is found in a moving study of an elder care home in Chicago by Tim Diamond (1992). His book *Making Gray Gold* is based on 18 months he spent training and then working as a nursing assistant. It is not only revealing about the ways in which gender identities are constructed through work, but also on the interconnections between gender and ethnicity in intimate and caring work in a private 'home' in Chicago.

Diamond is a sociologist who in the 1980s, as he was teaching and writing about medical sociology and the expansion of employment in the care sector, realized how little he knew about what life and work in care homes actually involved. After taking a part-time course, he worked for between three and four months at three different nursing homes. Committed to the methodology of participant observation, Diamond became an employee, doing exactly the same work as the other nursing assistants. As he started his research, he also read feminist literatures, especially about methodologies, believing that 'as a white man who wanted to work in this field, it might be valuable for me to experience some of the work that is done largely by women' (p. 6). His key theoretical influence was the US sociologist Dorothy Smith (1987, 1990, 1991), who argued powerfully for a method of practical research based on a focus on everyday work-tasks in the labour market as a way to build insights into how organizations and the wider society work. Diamond, after Smith, defines his work as an institutional ethnography which focuses on the actual daily social relations between individuals, rather than analysing the framework of texts, regulations and instructions within which care assistants must perform their work. As he notes, 'this standpoint is rendered invisible by the way most administrative and professional documents and texts are constructed' (Diamond 1992: 6). His aim was to make the invisible visible, showing the disjunctures that exist between teaching texts and training manuals and the nature of the work that care assistants undertake. Through his participation in the daily work in the institutions, he showed how what were defined as straightforward, routine manual tasks involved a great deal of caring emotional work by the low-paid and low-status workers, as well as the development of strong bonds between workers and the people they looked after.

While Diamond's approach is certainly ethnographic, because he is a white, middle-class man working in a position typically filled in Chicago by working-class women of colour, his study cannot be classified as an analytical auto-ethnography as defined by Anderson (2006). Diamond's outsider status – his skin colour, class and gender – meant that he could never become a complete member of the group that he was studying. Indeed, suspicion about his motives brought one job interview to an abrupt end with the question: 'Now why would a white guy want to work for these kinds of wages?' (Diamond 1992: 9), as Diamond was shown to the door. Right from the start, Diamond's gender was an issue – he was one of only two men on the training course, and because he was so out of place he was never able to benefit from the beneficial effects that Simpson (2004) identified for male nurses. Unlike nursing, being a nursing assistant in a care home was not constructed as a career and there was little opportunity for promotion, unless care workers trained as registered nurses. The work itself was also constructed as depending on essentialized feminine traits, despite the emphasis on acquiring specialist knowledge in the initial training course. As one of Diamond's co-students noted, 'What this work is going to take is a lot of mother's wit' (Diamond 1992: 17), by which she meant maternal feelings and skills. One of the class instructors defined it as 'a certain kind of just being there' (p. 18). But, as became clear to Diamond, just being there depended on maternal experience: an advantage that he just did not have. Here, Erma Douglas, a nursing assistant, is advising Diamond on washing people:

> I never wash the head when it's cold, and most times don't put soap on the face at all – it always gets in their eyes or mouth. She stared at me after this instruction, surprised that I did not already know this. 'You ain't had no babies, have you?' (Diamond 1992: 19)

Race, ethnicity and skin colour was a constant issue in the nursing homes where Diamond worked. The staff members were mostly people of colour, the residents in the main white. As Diamond explains:

> During the course of the research I worked with women, and a few men, from Haiti, the West Indies, Jamaica, Ghana, Nigeria, Mexico, Puerto Rico, India, South Korea, China and many from the Philippines. Never before, or since, have I been so acutely self-conscious of being a white American man. (Diamond 1992: 39)

In the early stages of his work, Diamond felt uneasy. The residents stared at him and often commented on his maleness, saying they reminded him of a nephew or a son, sometimes a doctor, but as time passed he realized that it was because he was atypical and so stood out as different: 'except for the few male residents and the occasional visitor, I was the only white man many would see from one end of the month to the next' (p. 39).

Some of the elderly residents, on occasion, exhibited racist attitudes. During the training, students were warned that 'patients have to be one size and one colour. Even if they tell you they want a white nurse instead of a black one, you have to swallow your pride and keep going' (p. 24). The course instructors were professional nurses and were predominantly white; the students non-white and despite the veneer of professionalism that overlay the teaching, it was clear to the trainees that they 'were not being trained to be professional nurses; we were being prepared for a different and lower stratum, in which most of our colleagues were non-white' (p. 28) – that sub-society identified by Bates (1993). A question about the function of the skin gave rise to a difficult moment in Diamond's group during the training period. When the correct answer emerged – to protect the body – the murmurings ceased but as Diamond noted, its role in constructing divisions of labour in the healthcare industry might have been addressed too (Diamond 1992: 30). And interestingly, and sadly, Diamond also noted that caring – as an emotional experience – was never mentioned either. The course was based on issues of germs and disease – the mind-body divide was at its centre and the body work part of caring, rather than emotional work, was the focus.

Nursing assistants made up three-quarters of the workforce at the homes were Diamond worked. Their work was hard, hours were long, their status and pay rates low. Many women had to meet rents that were up to half their take-home pay, and for women supporting dependent children, their wages typically fell below the official poverty line even though they were in full-time employment. Some of them did double shifts, or took a second caring job, often in a private home, to make ends meet, as Diamond discovered through casual talk at work:

A Filipino man remarked as we passed out trays together that this place had more hustle than the other place where he worked, on the evening shift. And then he turned to me to inquire 'where else do you work?'...

Two days later I was having lunch with Solange Ferier from Haiti. 'You know, I have done this job for six years in my country. There's one thing I learned when I came to the United States. Here you can't make it on just one job.' She tilted her head, looked at me curiously, then asked, 'You know, Tim, there's just one thing I don't understand about you. How do you make it on just one job?'

'Oh, ah ... I ... ah ... do a little teaching and tutoring when I get the chance' was my fumbling response. (Diamond 1992: 47)

The stories that Diamond tells in his book are moving and provide an extremely insightful account of the working conditions in nursing homes.

Yet, as I read the book, an underlying current of anxiety troubled me. Did he/should he tell his co-workers that he was an academic doing research for a book? Is it ethical to do covert research? The question clearly troubled Diamond himself and he directly addresses it in the introduction (p. 8). But even here his answer is not clear, as he reports that he told some people – his co-trainees for example and some of his co-workers, even a number of residents – but he also notes that 'as the study proceeded it was forced increasingly to become a piece of undercover research' (p. 8). The relationship between him and his co-workers and patients would have become different and presumably more problematic if they had not believed that he actually was, and so responded to him as, a care worker.

One of the key themes in Diamond's study is the contradiction between the ways in which care work is organized as a discrete series of tasks and the needs of the residents for more holistic caring that meets their emotional as well as physical requirements. The work is typically divided into a series of timed tasks – waking people, making beds, serving breakfast whether or not the residents wanted it then – leaving little room for personalized care. The timed daily schedules of tasks also excluded the less easily quantifiable parts of caring work, as Diamond explains here:

> The official tasks were difficult, sometime unpleasant and took some skill. But there was also a host of unspoken, unnamed demands before, during and after the tasks that presented problems, both physical and emotional. If the orders from the rational plan had parcelled out the tasks into a time-motion calculus that made sense in the abstract, carrying out the orders continually came up against the unplanned, fluid and contingent nature of everyday caring. (Diamond 1992: 143)

Meeting unplanned needs took experience and Diamond found that when the nursing assistants with whom he worked described the nature of their jobs it was often in terms of gaining experience or skills in getting to know people, in anticipating their needs and desires and seldom in terms of acquiring technical skills. The assistants spoke with pride and dignity about meeting the needs of the elderly people in their care, but also were aware of the necessity of maintaining the balance of 'getting close enough but not too close' (p. 148). As one colleague told Diamond, 'When you walk out of the room, you've got to leave them there and move on to someone else. You have got to practise hallway amnesia' (p. 159). The training and the timed daily work schedules made Diamond realize that 'formally the nursing assistant's job had nothing to do with talking' to patients (p. 159): instead, the work was constructed as and measured through a series of specific tasks. 'The social and emotional work was distilled into measures of productivity, and a responsive job was made over into a prescribed set of tasks' (p. 166).

In Diamond's and Twigg's case studies, as well as in Menzies Lyth's earlier work, the contradictions between rational, bureaucratic forms of organization and the provision of care are clear. These contradictions have grown. In the new millennium, there is an increasing emphasis in care work in all its settings – whether in the home, in elder care facilities or nursing homes and hospitals – on a rational model of timed tasks. The consequent work speed-up has increased pressure on workers. In British hospitals, for example, the stress involved in nursing increased significantly in the last years of the twentieth century and into the new millennium. Ackroyd and Bolton (1999) looked at the workload in the provision of gynaecological services in an NHS hospital in the 1990s and showed that although control and surveillance of the nurses had not increased, the numbers of staff had declined as the inflow and outflow of patients increased. Thus, while nurses still had a degree of autonomy, their workloads and work effort had increased significantly. As most nurses – like the care workers Twigg and Diamond worked with – have a commitment to their patients' welfare, there was little resistance to the increased effort needed, but growing dissatisfaction with a perceived inability to provide a good service as nurses felt they had to rush through their allocated workload. Rationalization and reorganization might seem to increase efficiency in public hospitals and increase profits in private facilities, but often at a cost which is borne by nurses and other care workers as their commitment to the emotional work of caring is a key part of their ethical stance. And yet, because of the associations between emotions, love and caring, these forms of work are poorly paid and often low status. Care work is, in the main, undertaken by women, and in the USA, but increasingly in the UK, by a migrant workforce. As Diamond comments on the expanding workforce in nursing homes in the US:

> The labour foundation of this developing institution is made up of workers drawn not just from within the society. The Filipino nursing labour force, as the primary example, has become central to the infrastructure of health services. The United States depends upon the Philippines not just for military bases in the Pacific but for its health care personnel as well. (Diamond 1992: 187)

A Filipino nursing force is also a significant part of provision in Canada (Kelly and Moya 2006) and the UK, where they are joined by a diverse set of workers from a growing number of nations. Among nurses in the UK, Smith and MacKintosh (2007) have identified an emerging ethnic hierarchy in which women of colour are found in the lower ranks of the profession and in less prestigious areas of work. As in the less credentialized areas of care work discussed above, the construction of the qualified nursing labour force in the UK is also dependent on foreign-born and/or foreign trained nurses. In the UK as a whole in 2003 about 3 per cent of all nurses were foreign trained, rising to 13 per cent in London (Ball and Pike 2004). In a survey of three inner London health trusts, the proportion of foreign-born nurses was found to be higher: 12–25 per cent (Buchan, Jobanputra and Gough 2004), drawn in order of significance from the Philippines, Australasia, India, Ghana and Nigeria. The nurses were concentrated in the basic levels – in grades D and E – and so had a lower average earning capacity than British-born nurses. These findings of an ethnic division of labour support the arguments in chapter 3 where I drew on Butler, Foucault, Ong and Bourdieu to show theoretically how hierarchies are constructed in the labour market on the basis of skin colour, ethnicity, class and gender. The embodied characteristics of potential workers, in combination with stereotypes about the suitability of differently endowed bodies held by employers, co-workers and the wider public, systematically disadvantage all those 'Others' who do not conform to hegemonic assumptions about the suitability of workers for embodied labour and care work.

Maintaining the Perfect Body

In the second part of this chapter, the focus shifts from elderly and sick bodies to an exploration of some of the jobs and occupations that are associated with bodily improvement, bodily adornment and beautification, including therapeutic massage, beauty parlours and hair salons and workers in gyms or as personal trainers. As Bryan Turner argued, the body is perhaps the last frontier in society over which the individual has (relative) control and so is a particular source of both anxiety and modification. One of the most significant ways in which bodies are modified is through surgery, where the change is permanent and often not reversible. I have chosen not to discuss cosmetic surgery here (there are several good discussions available, including Morgan 1994; Davis 1995, 2002; Gimlin 2002), but rather to focus on less radical forms of bodily modification, including diet, exercise and beautification. These modifications are based on active forms of self-modification in which individuals strive to conform to the regulatory norms that define the ideal body. This notion of a regulatory norm or ideal body parallels Judith Butler's arguments about compulsory heterosexuality introduced in chapter 3. Further, the bodily modifications desired are a clear example of Foucault's notion of disciplinary practices based on regulatory norms and capillary power. Foucault's analysis of disciplinary practices and forms of body power, what he termed the 'technologies of the self', is provocative and useful in analysing the different forms of self-disciplining body work. Many scholars have drawn on reinterpretations of Foucault's arguments (see, for example, Bartky 1993; Bordo 1993; Monaghan 2002) to help think through the meaning of body work. Through self-discipline, then, individuals strive to produce a modified self. In many cases, as these modifications typically demand considerable will power, the disciplinary practices are mediated or overseen by a specialist – a therapist, a dietician or a trainer, whether in a gym or other specialist space or in the client's home.

The focus of the last part of this chapter is on these mediated relationships, as it is here that interactive body work is located. The client purchases the support, the example and the services of another to aid in the desired bodily transformation. Exercise in the specialist spaces of the gym is perhaps one of the clearest examples of such body work, as here both customers and clients engage in body work on *themselves*. The desired transformations are thus a product of the *joint work* of the service purchaser and the service provider. In a study of why people work out in gyms and health clubs, Nick Crossley (2006) draws on both Foucault and Bourdieu to provide a framework for thinking about body work to achieve self-improvement. However, he did not consider the labour process but only the views of clients. Bourdieu (1977, 1984) himself did not devote much attention to body work, although he briefly suggested that subjects, especially middle-aged women, 'invest' in their bodies, perform work with and on them, in the expectation of returns to their investment in, for example, improved career prospects and in the relationship 'market'. The more recent emphasis on health and fitness and on fighting the effects of ageing might be added, as well as expanding the analysis beyond middle-aged women.

In comparison to 'working out', in other forms of transformative body work, the transformation is not only less permanent but also less active as the body of the client is passive – in the hair salon or beauty parlour, for example. Here intimate work on the body involving touching, cleansing, plucking and pummelling occurs, but the body of the client is typically supine or passive and not actively involved in the body work. This second type of transformative body work is perhaps closer to the patient being operated on in a hospital or the dead body in the funeral parlour being prepared to be viewed after death. In all cases the same sorts of issues about bodily boundaries, intimacy, privacy, touch and sexuality, odours, smells, disgust and visceral emotions that were explored in the earlier part of the chapter are also part of the social construction of the work of adornment and bodily modification and part of the interactive exchange between workers and clients.

The case study that completes this chapter is an example of passive body work: the working practices of beauty therapists. Here, men are out of place, especially if these workplaces are compared to working in gyms. Beauty therapy is yet another example of a feminized occupation. Not all self-improvement work, of course, is so feminized. Working as a trainer in a gym, for example, is a less gendered occupation and is open to interpretation as a gender-congruent form of work by both men and women. Think back to the work of Waquant that was introduced in chapter 6 and the masculinity of heavy, sweaty training as part of 'body-building' in a gym is clear, whereas a different set of associations with the female body, with supple and slender femininity produced by a 'work-out' in a health club, with aesthetics, as much as fitness, also allows the construction of this work as feminized. But in the close, confined and perfumed spaces of beauty parlours, as I explore below, men are redundant.

Working in Beauty Parlours: Physical and Emotional Work

There is a wide range of work under the general heading of beauty 'therapy', from hairdressing through massage to alternative or complementary treatments with parallels to medicine, where a degree of professionalization and regulation is now becoming more common. Osteopaths and chiropractors have official recognition in the UK and so are regulated; homeopaths, reflexologists, acupuncturists and herbalists are not, but draw on a professional rhetoric, whereas beauty therapists merge into the more purely decorative end of beauty treatments. Here I draw on the work of Paula Black and Ursula Sharma (2001), who interviewed eight beauty therapists employed by a department of a local college in a northern city and seven experienced therapists and salon owners in this and another city in the same part of the UK. They also visited a limited number of salons as clients.

When Sharma and Black (2001) undertook their empirical work in the late 1990s, the beauty industry was expanding. At the time there were over 7,000 outlets providing beauty therapies, employing 10,440 therapists (Beauty Industry Survey 1999). The workers were almost all women and their clientele was also predominantly female. Not unexpectedly, the rates of pay are generally low, just above the national minimum wage. Most salons were small and had only a few employees, but increasingly they are located within other spaces – department stores, hotels and health clubs, for example. There is also a movement to provide beauty treatments within hospitals to boost the spirits of long-term patients as well as to advise patients with disfiguring conditions.

The range of 'treatments' offered straddles the impermanent to the more permanent, involving varying degrees of technical competence. Sharma and Black (2001) list, for example, makeup and nail extensions as impermanent and eyebrow shaping/tinting, hair removal by electrolysis or waxing and slimming treatments as more permanent procedures. Therapists also increasingly practise what they term 'healing modalities such as aromatherapy, reflexology, massage of various kinds and occasionally Reiki (complementary therapies which claim to increase well-being and relieve stress rather than cure particular ailments)' (Sharma and Black 2001: 915). In this sense then, their work is more than merely cosmetic. Rather, it overlaps with the medical profession. As a respondent explained:

> I would say that it does offer a very high therapeutic angle to it; it boosts people's confidence and self-esteem … if somebody said years ago 'I am going to a beauty therapist's', you instantly thought beauty and vanity, the two go together. But these days there's more to salons, there's more emphasis on stress-related problems and that. (Black and Sharma 2001: 106)

In the interviews, the therapists talked about their work as playing a key part in making women feel better about themselves:

> You give them [a] feeling of well-being. I think they get that attention for that period of time which perhaps they don't get in other areas of their lives. (Sharma and Black 2001: 918)

Indeed, they emphasized this emotional part of their labours rather than the physical work of improving women's looks. They defined beauty therapy as work involved with feelings as well as with the body and insisted it was more than decoration. In part this emphasis was to counter the general public's view of beauty therapy as trivial or as akin to sex work:

> We all have to deal with the 'oh, they just paint nails and do massage – nudge, nudge, wink, wink!' Men tend to regard it smuttily and women … tend to think we are all brainless bimbos. (Sharma and Black 2001: 917)

One of the ways to counter these stereotypes is to emphasize the associations with medical treatments by, for example, presenting a clinical appearance through wearing clean white overalls which de-emphasizes the worker's sexuality and emphasizes her professionalism. Therapists also explained that they took care when taking on male clients, especially for treatments such as waxing body hair. The growing male market for such treatments has made the need to de-sexualize therapists' appearance and persona even more significant.

The therapists defined the work they did as pampering, treating and grooming, but also emphasized that all three sets of tasks demanded an emotional involvement and relationship with each client as an individual. The therapist uses her personality and style to build up an individualized relationship with each of her clients. As one women explained, 'you get clients who are sort of *your* clients' (Sharm and Black 2001: 919). Therapists build these connections by gauging the mood of their clients, whether they want to talk during a treatment, for example. They must also learn to manage their own emotions and to steer clear of controversial topics. One therapist explained the informal set of rules that operated in her salon:

> We have, like, these rules. You don't talk about religion, you don't talk about sex, you don't talk about politics; you have got to know your key things you do not discuss. But then you have still got to be quite relaxed with that client ... you've got to show you're a human being as well, and if they want to talk about anything then, you know, you are there to listen. (Sharma and Black 2001: 920)

Beauty therapists, like the nurses and care workers already discussed, also have to deal with intimate contact with clients' bodies. They have to negotiate a contested relationship in which the clients may feel awkward and embarrassed that they do not match the norms of feminine beauty promoted in the salon and more widely. Workers, too, have to cope with their own feelings of distaste or revulsion. They may have to disguise feelings of boredom or distaste as well as ensure they do not burden clients with their own emotional issues, but the performance of emotional labour is not necessarily a sham. Their professionalism 'may demand self-control and deference to a client's mood and wishes' (p. 921), but a degree of reciprocity is developed with some clients. The close, personal and repeated interaction enables a different sort of relationship to be established than, for example, the more ephemeral relationships that occur between the workers in fast food outlets and their customers that are the subject of chapter 8. In this way, the work of beauty therapists is more similar to that of the care workers, as both groups find pleasure and job satisfaction in establishing emotional ties with their clients. Sharma and Black emphasized this sense of satisfaction reported by many therapists, contrasting their attitude with those of the airline workers interviewed by Hochschild (1983). She found high levels of dissatisfaction among these workers, in part explained by close monitoring of their bodily performance. Both the domiciliary care workers and the beauty therapists, especially those working in small salons, had a greater degree of independence in their workplaces.

Beauty therapy is an exceptionally gendered occupation that takes place in a set of spaces also associated with women. It not only relies on feminine attributes such as tact, sensitivity and compassion, but it also works on the female body and its adornment. Although as I have noted, workers present a non-sexualized appearance, they must also conform to acceptable versions of heterosexual attractiveness, wearing visible make up, for example, and having a neat but flattering hairstyle. They must continually work on their own bodies, as well as those of their clients, to meet these expectations. Men's entry to this predominantly female world raises similar questions about how to fit into or challenge the sets of associations between this type of work and femininity, as in the case of nursing explored earlier. While there is scope for men to redefine various aspects of the therapeutic role as congruent with masculinity, through links with psychoanalysis or other branches of medicine, for example, it is less obvious how they might reconstruct some of the more intimate body work tasks like plucking and dyeing bodily hair unless, like male care workers, they are restricted to practising body work on other men. The classic example is the barber's shop, where men cut other men's hair in an atmosphere and space that is typically more redolent of utilitarianism than pampering. For some years, my own partner had his hair cut at a barber's that advertised itself as 'Mick's: no waiting'. This seems to me to sum up the difference between (most) men and (most) women's attitudes towards bodily adornment. Whenever I visit 'health clubs' in hotels, I look, usually in vain, for male beauty therapists but see only men as personal trainers, men working in the gym, men as life guards by pools and men as bar staff in the leisure spaces.

Men are out of place in the perfumed worlds of pampering, treatment and grooming described by Black and Sharma, just as Diamond was in the predominantly female worlds of Chicago nursing homes. Beauty parlours are an even more extreme example of a feminized world than care homes, in which the joint endeavours of the female worker and female customers reproduce as near as is possible an idealized image of hegemonic femininity.

In the next chapter, I continue this theme of men as out of place in an exploration of the ways in which young working-class men's sense of themselves as masculine disadvantages them in other types of low tech consumer service sector occupations: in hospitality and retail jobs that are often the only types of work easily available to young men with few skills and credentials. In these jobs the interactive aspects of service work are more significant than the body work analysed here, although workers' own embodied performances are once again a key part of the employment contract and the service exchange.

Plate 5 Learning to mop: the acquisition of 'feminine' skills in a diner. Photo © Plush Studios/Getty Images.

8

Warm Bodies: Doing Deference in Routine Interactive Work

> It is humbling, this business of applying for low-wage jobs, consisting as it does
> of offering yourself – your energy, your smile, your real or faked lifetime experi-
> ence – to a series of people for whom this is just not a very interesting
> package.
>
> Barbara Ehrenreich, *Nickel and Dimed*, 2001

Most of the forms of body work discussed so far have included the intimate touching of the bodies of others, involving pleasure and pain, as well as various acts of caring for the bodies of others. While not necessarily including touch, interactive service exchanges with physically present clients are also an expanding part of the consumer economy. Here the physical production of a particular appearance and the deferential performance of an often-scripted exchange are an essential part of embodied social relations and the construction of identity in the workplace. In retail sales, in leisure industries and in hospitality, both employers and customers have expectations of appropriate forms of servicing their needs by bodies that are present during the exchange. A range of expectations of and attitudes about 'good service' by appropriately embodied workers thus affects recruitment and employment practices. Customers' expectations and evaluations of the service provision influence their propensity to return and repurchase the service and so materially affect profit levels. As these types of services typically are used up in the exchange, there is an almost infinite possibility for reprovision if the service is deemed acceptable.

In many forms of interactive service exchanges, the heterosexual matrix, defined by Butler, operates to associate deference, servility and an appropriate degree of 'invisibility' (that is, servicing without being noticed) with differently raced, classed and gendered bodies. As I explore here, masculinity is often, if not inevitably, a disadvantage in interactive service employment at the bottom end of the labour market. Hegemonic versions of masculinity typically are associated with dominance and action, 'in your face' presence, rather than with deference and empathetic forms of low-key serving and servicing. This chapter explores the connections between masculine performativity, unskilled, poorly educated, working-class young men and access to low-waged occupations in interactive services. Although the main emphasis is on masculinity, I also explore the ways in which ethnicity, nationality and skin colour intersect to produce a hierarchy of 'appropriate' bodies for different positions in service occupations, drawing on the theoretical analysis of complex inequality that I outlined in chapter 3.

I explore the world of interactive service employment for young men in two British cities – Sheffield and Cambridge – and in New York City, as they look for their first jobs. These cities are places where well-paid opportunities in manufacturing employment have disappeared and jobs in fast food and retail outlets and in the hospitality sector are often the only options for many of the localities' youth. In the British cities, the young men at the centre of the chapter are white and working class, often long-term residents, in the main living in intact nuclear families. In NYC the young people are more obviously disadvantaged in the labour market: many of them are living in single-parent families or are already parents themselves and most of them are Black. The British example is from my own empirical research, undertaken between 2001 and 2002 (McDowell 2003). The NYC example is based on work by Philippe Bourgois (1995) and Katherine Newman (1999), both anthropologists, carried out in the 1990s. More recently, Newman (2006) traced the respondents in her earlier NYC study across a range of US labour markets to see whether they had been able to escape bottom-end work and poverty wages. Her study is an excellent, and uncommon, example of how a detailed case study may become a larger-scale analysis of the labour market through longitudinal analysis. Problems of cost as well as tracing people often preclude this type of follow-up study.

In the second part of the chapter, the focus shifts from the social construction of masculinity by fast food workers to hotel workers: the warm bodies of the chapter's title. Here workers are recruited for poorly paid vacancies in both front- and back-stage areas of hotels. I draw on my study of a west London hotel that employs recent migrants to the UK, both young men and young women, as waiters, kitchen assistants and room cleaners (McDowell, Batnizky and Dyer 2007; McDowell 2008). I show how ethnicity, gender, class and nationality intersect to construct different workers as more or less suitable for particular kinds of work, creating a hierarchy of acceptability. Employment agencies play an increasingly significant role in assembling transnational workers such as these across an expanding spatial scale, even for the most low-paid types of work. Like childcare and sex workers of earlier chapters, transnational workers also perform many of the most local and place-specific jobs in the hospitality industry.

What is Happening to Young Men in the Consumer Service Economy?

As I argued in chapter 3, the social construction of identity in the labour market is a work in progress: men and women 'do gender' in different ways in the spaces and places of their everyday lives as well as within the context of those institutions that regulate their behaviour. Although masculinity and femininity are always contingent and actively constructed, historically and culturally situated and multiple, nevertheless, dominant or hegemonic versions of identity are identifiable. 'Hegemonic masculinity is the standard bearer of what it means to be a "real" man or boy and many males draw inspiration from its cultural library of resources' (Kenway and Fitzclarence 1997: 118). For working-class men, hegemonic masculinity is, or was, associated with a lifetime's labour market participation and with the acceptance of responsibility for dependents, despite the growing challenge from women's increased economic participation. Hard manual labour during the working week and uninhibited leisure and consumption activities in non-working hours, particularly by young men, was common, especially in regions dominated by heavy industry. For working-class men, part of an acceptable version of masculinity lies (or used to lie) in the associations between attributes of embodied physicality and the public arenas of the workplace, as well as in the streets and in leisure spaces. Compared to the highly valued cerebral rationality of middle-class masculinity, working-class men's only advantage in the labour market lay in their physical strength (Connell 1995, 2000), as studies of traditional occupations such as mining, fishing and heavy manufacturing industries have made clear (Samuel 1977). In a classic study of young men at the point of leaving school for employment, Paul Willis (1977) demonstrated how in the postwar decades the British school system reinforced these associations by socializing working-class boys in ways that both prepared them for and restricted them to manual labour. Thus the patterns of work that typified the long Fordist era between the end of the Second World War and the 1970s were reinforced by social institutions such as schools, the family and the housing market, producing a stable gender regime in which 'real' men did 'men's work'.

The growing dominance of interactive forms of employment has recast this gender regime (Connell 1995; Walby 1990). As growing numbers of women enter the social relations of waged work and young women increasingly gain educational credentials that equip them for social and occupational mobility, young men, but particularly working-class young men, with little educational and social capital, are disadvantaged in a labour market that valorizes deference and civility. Working-class young men, many of whose fathers had worked in manufacturing employment, face a set of labour market opportunities predominantly restricted to poorly paid servicing jobs. For prospective employers, the stroppy, macho, often awkward young men that they interview to fill vacancies are less appealing prospects than either young women from the same class position or the growing number of older women returning to work as their children go to school (Berthoud 2007), often to make up for the declining incomes of their husbands and partners who are also affected by declining opportunities for men (Alcock et al. 2003). If we add to this prospective labour force growing numbers of schoolchildren and students in higher education – the latter adversely affected by rising fees and inadequate loans (Christie, Munro and Rettig 2001) – the position for young, unskilled, male school leavers looks increasingly unpromising, especially when unemployment is rising.

For many young working-class men, masculinity, as it is conventionally constructed and enacted, is now a disadvantage rather than an advantage as they leave school and seek to enter the labour market. At school, working-class boys find it hard to conform to the demands of a system that depends in large part on the completion of tasks coded as compliant or feminine – the production of course work to an agreed timetable and standards of neatness that seem difficult for many boys to achieve (Willis 2000; Jackson 2002; Reay 2002; Arnot 2004). In the classroom, their behaviour is often too boisterous and so they may be excluded, missing key parts of their education. On leaving school, typically at the earliest official leaving age (currently 16 in the UK), they find that they are under-educated and unqualified, ineligible not only for high-tech, high-status occupations in the service sector but also for many of the expanding high-touch jobs, such as nursing or care work. Restricted to the types of deferential interactive servicing work available in shops, restaurants and fast food outlets, or possibly in office services, young working-class men find themselves as unqualified and inappropriately embodied in their search for work. In these latter forms of interactive work, success is dependent on being able to produce 'service with a smile', the ability and willingness to defer to clients and superiors, and the presentation of a neat, clean, servile body that conforms to a set of organizational norms in terms of weight, height, accent, hirsuteness and decoration.

Iris Young, in her definition of cultural oppression, included the old and overweight and the physically unattractive among the range of people disadvantaged as potential employees in consumer services. More surprisingly, fit and healthy young men may now be counted among the culturally oppressed, as their embodiment, their looks, their stance, their embodied hexis, seem threatening to potential employers and customers. There is a long history of the association of young working-class men with threats and fear of the mob, demonstrating that the disadvantages of youthful masculinity are not a new phenomenon (Cohen 1973; Pearson 1983). What is new is the growing disjuncture between the characteristics of masculinity and the requirements of bottom-end, high-touch, interactive employment. As Bourdieu (1984: 96) argued, the construction of a version of tough, aggressive masculinity has long been

> just one of the ways of making a virtue out of necessity. The manifestation of an unreasoning commitment to realism and cynicism, the rejection of feeling and sensitivity identified with feminine sentimentality and effeminacy, the obligation to be tough with oneself and others ... are a way of resigning one-self to a world with no way out, dominated by poverty and the law of the jungle, discrimination and violence, where morality and sensitivity bring no benefit whatsoever.

While Bourdieu's claim explains young men's behaviour in the tough streets of the inner areas of deindustrializing cities or the global cities dominated by both producer and consumer services, this set of attitudes – the tough, devil-may-care approach – is exactly what disqualifies them from many of the jobs at the bottom end of the service sector where 'feminine sentimentality' is a valued attribute.

Humiliation and respect in NYC: interpersonal subordination in service work

The ways in which the version of streetwise masculinity identified by Bourdieu disqualifies young men from labour market opportunities in the service economy have been vividly illustrated in a study of young minority men (where skin colour is an additional source of fear and loathing of working-class youth) in a New York City neighbourhood disadvantaged by the almost complete disappearance of blue collar jobs. Philippe Bourgois (1995) lived in his research area and 'hung out' with the men on the streets in a restricted form of auto-ethnography, refraining from participation in drug dealing, although he was sometimes arrested on suspicion and had a hard time convincing the NYPD that he was an academic. He found that the particular version of a tough, aggressive, sexualized street credibility valorized by young men in that locality disqualified them from the only types of vacancies then expanding in NYC. As he noted, these youths 'find themselves propelled headlong into an explosive confrontation between their sense of cultural dignity versus the humiliating interpersonal subordination of service work' (Bourgois 1995: 14). Drawing on Bourdieu, he argued that the social and cultural capital accumulated by working-class youths on the streets in NYC was inappropriate and so unvalued in white collar workplaces:

> Their interpersonal social skills are even more inadequate than their limited professional capacities. They do not know how to look at their fellow service workers – let alone their supervisors – without intimidating them. They cannot

walk down the hallway to the water fountain without unconsciously swaying their shoulders aggressively as if patrolling their home turf. Gender barriers are an even more culturally charged realm. They are repeatedly reprimanded for offending co-workers with sexually aggressive behaviour. (pp. 142–3)

In a study in the UK, I looked at the employment prospects of a similar group of poorly educated, working-class young men, although they were white (McDowell 2003). In both Cambridge and Sheffield, part-time work, non-standard or casual employment in 'bad jobs' (McGovern, Smeaton and Hill 2004) or forms of self-employment were the only options for most working-class youths on leaving school. Like Bourgois, I found that not only did the limited educational and cultural capital of these men disqualify them from many of the vacancies on offer, but so too did their aspirations. Many young men to whom I talked had clear views about the types of work they were prepared to consider, regarding most routine, casualized service sector work as 'women's work' and so beneath their dignity. Even if they were prepared to consider employment in the shops, clubs and fast food outlets that were the main source of work for unqualified school leavers in both cities, they often disqualified themselves as potential employees by their appearance (piercings and tattoos as well as inappropriate clothes) and their attitudes during the recruitment process. Employers read the surface signals of bodily demeanour, dress and language as indicators of unsuitability for the vacancies on offer. If these young men were successful in finding employment, then their sexualized, aggressive embodied interactions, especially with women co-workers and superiors, often disqualified them after a week or two. Many of them found it hard to perform the deferential servility required in the service economy.

But not all of them found themselves disqualified, whether by employers or their own choice. Some of the school leavers to whom I talked were prepared to knuckle down and produce the sort of deferential performance and servile docility essential to holding on to their jobs. For these young men, a common way of justifying their employment was to locate it within a conventional narrative about how masculine identity involved participation in waged work and the ability through this to provide for others. Through the attribution of masculine characteristics to particular types of work, these young men also challenged the image of service employment as necessarily deferential and demeaning. While serving in shops was often seen as both servile and feminized, if the work involved selling sports goods, for example, the young men were able to translate it into giving advice to sportsmen and so appropriately gendered as masculine. This parallels Leidner's (1993) findings in the insurance industry. Women sellers talked about empathy as a key attribute of success, whereas men in the same job reinterpreted it in a masculine frame, through a rhetoric of overcoming resistance and breaking down barriers and so congruent with the hegemonic construction of masculine identity.

The men in my study who were in employment found themselves involved in a complex negotiation between the sort of respectability required to perform satisfactorily in the workplace and the macho/laddish street credibility that brought respect from their peers. In some arenas of employment, this negotiation placed difficult demands on young men. For workers in fast food outlets, for example, dismissive attitudes of customers or the expectations of friends of free food or drink placed these men in an awkward contradiction, reducing their sense of independence. As Greg in Sheffield noted:

> Too many people I know from school come in here. And they expect free drinks and fries and all that. They come up to me talking loud. I hate it 'cos we are watched all the time, you know. I can't give stuff away.

Strangers are also hard to deal with, especially at peak times when queues are long and when, especially after clubs and bars have closed, customers may be inebriated or aggressive. Here is Greg again:

> I hate it when customers are rude to you. They sometimes speak to you like dirt, you know; I think they think they are better than us ... you want to lean across the counter and thump some of 'em, but I'd be sacked if I did, so it's not even worth thinking about.

Thus, the exchange with customers is a challenge to young people's sense of themselves, making it difficult to produce an acceptable performance despite the induction programmes that teach workers how to produce a 'service with a smile', even when customers are rude:

> Servicing the customer with a smile pleases management because making money depends on keeping the clientele happy, but it can be an exercise in humiliation for teenagers. It is hard for them to refrain from reading this public nastiness as another instance of society's low estimation of their worth. But they soon realize that if they want to hold on to their minimum wage jobs they have to tolerate comments that would almost certainly provoke a fistfight outside the workplace. (Newman 1999: 89)

The social characteristics of inner-city male youth culture, especially in its celebration of independence and rejection of deference, construct young men as singularly unsuited to this form of work. As Newman (1999: 95) argued, based on a study of fast food employment in New York City:

> Ghetto youth are particularly sensitive to the status degradation entailed in stigmatized employment.... [A] high premium is placed on independence,

autonomy and respect among minority youth in inner-city communities – particularly by young men. No small amount of mayhem is committed every year in the name of injured pride. Hence jobs that routinely demand displays of deference force those who hold them to violate 'macho' behaviour codes that are central to the definition of teen culture. There are, therefore, considerable social risks involved in seeking a fast food job in the first place, one that the employees and job-seekers are keenly aware of from the very beginning of their search for employment.

Thus these jobs, despite their low pay and status, are hard for young men to obtain and even harder to hold on to. In Burger Barn (a pseudonym) in Harlem where Newman undertook her research, the average job tenure was less than six months. The fast food chain McDonald's has an extremely high turnover of staff in both the UK and the USA, especially among entry level workers, although it does have a reasonable record of training programmes that allow workers to move up and acquire transferable skills once they demonstrate some commitment.

Entry level work is demanding as well as demeaning. Once a job is secured, working conditions not only challenge young people's sense of the respect owed to them, but also make physical demands on workers. Tasks are broken down into component parts, actions are timed and service has to be speedy. Workers need to have a high level of manual dexterity and the ability to work accurately at speed and under pressure. In this sense, producing the burger and fries wrapped and ready for sale is more like a Fordist manufacturing process than an interactive service exchange (see also Gabriel 1988). Here is Richard in Sheffield describing the way that a burger is made ready for sale:

> So to do like a Big Mac or something, there's one on table dresses t'Big Mac, passes the bun on to grill and another takes it off so its reet quick, like a formation 'cos it only takes 30 seconds to get like five or six Macs off.

This Taylorist team work is done under surveillance. Managers watch the team's performance, timing the tasks. Workers often find this intrusive and oppressive.

> They push people around, management does ... people often don't last long here. They leave after about two months because they can't handle it. They answer back and that.

As I argued in the last chapter, even though caring work is also often degrading and demeaning, domiciliary workers value their independence and the absence of everyday surveillance. For counter workers in fast food, the job is an embodied interactive exchange with little room for initiative or for constructing a personalized exchange.

Wearing a distinctive, often-constricting uniform is seen as a further humiliation. Danielle, for example, told Newman that she was proud of holding down a job but preferred her friends not to know where it was. As she explained:

> I ain't saying I am ashamed of my job, but I wouldn't walk down the street wearing the uniform ... guys know you work there will say 'Hi, Burger Barn'. I ain't gonna lie and say I am not ashamed, period. But I'm proud that I am working. (Newman 1999: 92)

As Ong (1991: 290) noted in research on factory workers in Malaysia, bodily regulation through prescribed clothing and footwear is often resented by workers as a bodily affront and a challenge to their sense of an autonomous self-identity. Young men find it particularly difficult to produce the required scripted deferential performance, especially in the face of perceived humiliation. Young women tend to be rather better at it, or more adept at pretence – the management of emotions defined by Hochschild (1983). Young women typically are socialized to present a modest demeanour and deferential attitudes to adult authority, although perhaps less so than in earlier decades, as well as being more able or willing to defuse difficult situations. In these situations, many young men simply leave their jobs in fast food and move on into other forms of casualized work in bars or clubs. For employers, these young workers are the generic workers defined by Castells (2000): unskilled, and entirely replaceable as each year another cohort of unqualified school leavers enters the job market. Nevertheless, as Newman argues, the young men and women who do manage to stick it out in the bottom end of the service sector deserve respect rather than dismissal by workers in higher-status jobs. Holding on to a job, albeit and especially one involving humiliating interpersonal interactions, represents strength of character and determination for young people who are among the most disadvantaged in the polarized labour market dominating service economies.

These case studies were carried out in cities that differ considerably in size, their labour markets and the range of jobs on offer, as well as in costs of living and alternative opportunities for work and leisure. Yet the young men at their centre live similar lives. They are socially and economically disadvantaged by their class position, their lack of skills and their masculinity, and geographically disadvantaged by their spatial immobility. In all three cities, the men who were interviewed seldom looked for work beyond their immediate neighbourhood, restricting their opportunities to the sorts of low-wage work in consumer services that were often all that was available in the locality. In the second part of this chapter, I look at the same types of work but instead of disadvantaged young men who were born and grew up in the city, I explore the lives of migrant workers in low-waged, interactive service employment in cities far from their place of birth.

Warm Bodies for Hotel Work

The focus in this section is on recruitment and employment practices in a different part of the hospitality industry – a hotel rather than fast food outlets. Here I explore how a large hotel in Greater London relies on an almost entirely non-British born labour force, typically recruited through – and in some parts of the hotel employed by – employment agencies, rather than directly by the hotel. The term 'warm bodies' is a (derogatory) term used by agencies to refer to supplies of unskilled workers for general menial work: a further example of Castell's (2001) generic workers. As the end employer, the hotel might, for example, ask an agency to supply 25 workers to clean rooms, with no further specification of their characteristics or attributes. All that is required are bodies willing to work: hence the term in the subtitle. However, through this and other recruitment practices, the hotel management assembles and regulates a segmented labour force. Through its reliance on foreign-born workers, hierarchies of gender, race and class are constructed, enacted and maintained (McDowell, Batnizky and Dyer 2007). Links between ethnicity, gender, class and migration – the intersectionality discussed in chapter 3 – and the stereotypical characteristics associated with these social attributes allow certain bodies to be constructed as suitable, and others inappropriate, for particular forms of menial or interactive work, as the empirical work described below makes clear.

This case study is a London-based one, the British city where ethnic diversity in the labour market is most significant (May et al. 2007). I have called the hotel where the research took place Bellman International (BI), a pseudonym. It is a branch of an international chain, well known for its aggressive employment practices and for its policy of non-unionization. BI is located in west London, near to Heathrow Airport, and it caters for the rich and famous as well as business travellers, the more affluent end of the holiday trade and a random selection of various others with business to transact in the vicinity of the airport. Its room rates are towards, though not right at the top end, of the London market and its clientele is international: guests from all parts of the world stay there. This diversity is matched by the hotel's employees. Individuals from almost thirty nationalities were employed in 2006, on a range of contracts and different terms and conditions. Less than 10 per cent of its total employees are British born.

Numerous studies of migrant workers have documented how newcomers, especially those with few educational credentials (or at least those recognized by the 'host' nation) or skills, are excluded from well-paid jobs and tend to crowd into bottom-end jobs in poorly paid sectors of the labour market (Piore 1979; Bailey and Waldinger 1991; Anderson 2001; Hondagneu-Sotelo 2001; Parreñas 2001; Rosewarne 2001; Ehrenreich and Hochschild 2003).

Class, skin colour, language problems and low levels of unionization (Milkman 2000) combine to reinforce the disadvantages new entrants to the labour market typically experience. Poor conditions, long or irregular hours, low pay and discriminatory promotion prospects combine to reinforce their initial segregation. Although hotels also employ professional workers, they have become an almost quintessential site for analyses of low-paid interactive work. Labour economists, geographers and sociologists interested in labour segregation and exploitation in service-based economies have turned to the hotel as an ideal site, as hotels are notorious employers of cheap, relatively docile and insecure migrant labour (Savage 1998; Waldinger and Lichter 2003; Seifert and Massing 2006; Tufts 2006; *Antipode* 2006). Thus through this site, the ways in which class, gender, national origin and skin colour interact in the construction of ideal workers for different types of interactive service work are revealed. Hotels are also particularly useful sites for understanding the connections between identity, interactive work and performativity, as issues of 'doing gender' and 'doing ethnicity' in forms of body work that depend on personal relationships between the seller and purchaser of a service are perhaps at their starkest here.

Hotels combine many of the features of the jobs and organizations examined in earlier chapters. Like hospitals, they cater for customers who are away from home, sleeping within the walls of what are workplaces for employees. They also (at least in up-market hotels) provide services for bodily adornment and exercise – hair and beauty salons, gyms and pools, for example. They include bars and restaurants, which are sometimes the location for the arrangement of sexual encounters, often commodified ones, as well as sites to purchase food and drink. Further, hotels are organizations that provide both visible (checking in, waiting on table) and invisible (cleaning rooms and corridors, providing clean linen) services in which an often-sexualized 'service with a smile' is highly valued, if not highly rewarded.

Managers, employees and guests construct their identities and forge social relationships in the interactions and exchanges that take place in the front and back-stage arenas of the hotel (Goffman 1959; Waldinger 1992; Guerrier and Adib 2000). Questions about gender, nationality, personal style, embodiment, skin colour, weight, bodily hygiene and language abilities (especially when there is an international client basis) are all crucial parts both of the decision to hire categorically distinctive workers to produce acceptable workplace performances that match the fantasies and desires not only of the employers but, more importantly, the guest. The aim is to produce a particular experience of 'hospitality' for guests in which they feel, at a minimum, valued and, at best, pampered. In hotels, as in a wide range of other service sector workplaces, gender remains a crucial discriminator of workplace acceptability. Men and women continue to do different types of work. In hotels 'when employers are looking for the most "appropriate" worker, suitability is largely determined categorically, heavily influenced by the sex of the person who typically fills the job' (Waldinger and Lichter 2003: 8). In a workplace which reproduces many of the relationships of the home – the provision of domestic comforts of food, rest and solace for the weary worker/traveller, albeit in semi-public spaces – gender matters. Thus, as sites for study, hotels embody many of the social relationships already dissected, providing an excellent final example.

As well as categorical associations – between gender and occupational segregation, for example – a set of stereotypical assumptions also influences employers' and managers' judgements about the acceptability of particular workplace presences and performances in the different spaces of a hotel. As I have already shown, in all workplaces, assumptions about appropriate and acceptable versions of femininity and masculinity, about the requirement of, say, docility and caring for nannies or fast food workers, physical strength for boxers and fire fighters, are part of the decision about who is able to perform particular jobs satisfactorily. For migrant workers, assumptions about national characteristics are also mapped onto the bodies of workers in deciding who is suitable to do what; examples include the ways in which assumptions about their 'natural' caring abilities are embodied in Filipina nursing assist-ants (Kelly and Moya 2006) or the apparently 'natural' maternal feelings of Black servants (Glenn 1992, 2001). These notions run through hiring deci-sions, as well as structuring the social relations between managers and employees, between workers and guests and between co-workers. They affect and are reflected in embodied performances that are regarded as more or less legitimate and so in the allocation of individuals with particular sets of social characteristics, embodied attributes and social and economic skills to different occupational positions in the hotel. The extreme division of labour in hotels in general and in BI in particular – by gender and national origin – illustrates the impact of these discourses of workers' suitability and unsuita-bility, as well as the extent to which employees both internalize and challenge these views in their daily working practices.

As I argued in chapter 3, the notion of interpellation (call and response) is a useful way to capture the ways in which workers are positioned within these discourses about suitability. In interactive service work in a hotel, cli-ents, customers and guests, as well as employers and managers, construct a series of imaginaries in anticipating the interactions that will take place within its walls. Workers have to 'look the part' in interactions with clients and their appearance and bodily style, as well as their performance, becomes a crucial part of the service. In the hospitality industry, customers' fantasies are often based on ideas of desire, if not actual sex, which may place work-ers, especially women, in a position that makes them vulnerable to sexually suggestive remarks and behaviour and other forms of harassment. Under new regulations (introduced in April 2008), British employers are required to protect their staff from sexual harassment by customers, suppliers and others encountered in the course of their work. Earlier legislation gave protection from harassment by co-workers. The legislation is important for the hotel and restaurant industry which employs 670,000 women in the UK and where, according to the Equal Opportunities Commission (now replaced by the Equality and Human Rights Commission), 'sexual harassment by customers is rife' (Dyer 2008). It is difficult, however, to see how these new rules will be effective in an industry that emphasizes customer satisfaction, requiring workers to produce gendered performances that emphasize sociability, deference and a particular version of courtesy.

In the rest of this chapter, I explore first how and why workers are recruited by BI and then how they enact an embodied workplace performance. In exploring performativity, I use the example of the greeting work undertaken at the front desk – the first site where clients are welcomed and reassured about the services available to them during their stay.

Who do managers employ and why?

Hotels in global cities typically rely on both a casualized and minority/migrant labour supply which brings advantages of low labour costs and flexible (or exploitative) employment contracts (Bernhardt, Dresser and Hatton 2003). Mason and Wilson (2003: 37), in a survey of employers in the hospitality sector in the UK, found that 'the trend to employment of foreign workers was accelerating', partly as a result of the EU's single labour market, and many establishments were heavily reliant upon foreign workers. These filled technical skill gaps (such as cooking and bar skills), but many employers also referred to the work attitudes of foreign workers as being more appropriate for the hospitality sector. Here we see an indication of the ways in which stereotypical assumptions might advantage foreign workers, at least in numerical terms, rather than low-skilled British workers, whose commitment to hard work was judged inadequate (Anderson, Ruhs and Spencer 2006). Whether surveys of employers are entirely believable is an important question. It is unlikely that employers would agree to suggestions that migrants, valuable as they are, are more easily exploitable or underpaid.

Lee-Anne, the director of human resources at BI, believes that British workers will not consider hotel work because it is too poorly paid:

> The way of the world, it's changed.... One big thing for hotels is, English people in this country will not work for the salary that is being offered within hotels, so you have to look elsewhere.

BI's workers were instead assembled and employed from a transnational pool of increasingly diverse workers, often through the use of employment agencies. Low-skilled jobs in the hotel were almost all filled by agency workers, employed on temporary contracts held by recruitment agencies in London. The hotel also used foreign-based agencies to recruit more skilled workers. Trainee managers, for example, all of them young men, were recruited in India. Twice-yearly trips were made to select potential employees whose visas would then be arranged by the hotel. The recruits initially entered the UK as student participants in a two-year management trainee programme, during which they were moved between sections to gain experience in different areas of hotel work. Typically, they started as waiters.

While this decision to recruit in India might seem surprising, young men from India and Bangladesh are often sourced by multinational tourism corporations, based both in India and elsewhere, to work in UK hotels. In 2002, for example, 35 per cent of all the work permits issued in the UK for employees in the hospitality sector went to Indian and Bangladeshi nationals, reflecting Britain's colonial legacy. Many of these work permits, however, were for low-skilled, often poorly educated employees in restaurants and cafés rather than in hotels. The BI employees were different: in the main well educated, all of them had post-school qualifications, and several had degrees.

Why did the managers of BI prefer to recruit overseas than from among, for example, the local Asian population that lives in west London? Here it seems a series of managerial assumptions coincided. These men were constructed as ambitious yet deferential, regarded as 'good management material' and, significantly, committed in most cases to return to India once fully trained. Anna, an international recruiter for the BI chain, suggested that the hotel focused on India on the advice of an agent in Delhi who argued: 'You know, you really should be going and seeing what India has to offer because hotels are fabulous, the service is fantastic and there's a very large number of people who would like to come and obviously work for a company like yours.' In an interview, Anna spelt out the characteristics that the firm was looking for, emphasizing the key significance of interpersonal interactive skills:

> We look for someone who's got a very strong aptitude to interact with customers because that's key, that's what hospitality is about, whether that be on the front desk, whether that be in the restaurant, whether that be conference banqueting, even housekeeping, it's really important.... We look for somebody who has a style basically, the kind of person that when you first meet, you'll warm to because that's the image that Bellman has. It's all about hospitality, and we also look for someone who's very well presented.

Here is evidence of the attributes valued by a high-end hotel chain: style and interactive skills, characteristics that are hard to quantify and open to interpretation by the recruitment team on the basis of a candidate's embodied performance in an interview. Since 2004, Anna noted, a number of countries in

Eastern Europe had been targeted as a potential place for direct recruitment. 'We went to Hungary first and then to Latvia, but so far it hasn't been very successful – not like the Indian recruitment.' Part of the reason, she suggested, was related to presentation of self:

> The actual aptitude I was talking about, the smiley, bubbly hospitality attitude, is not as prevalent in the people we interviewed, they're a lot more serious ... so there wasn't a natural what I call personality or that demonstration of 'I'm here for the customer', so that was a little bit of a concern.

As Goleman (1998: 3) argued in his assessment of the demands of work in the new economy, 'The rules for work are changing. We are being judged by a new yardstick: not just how smart we are, or by our training or expertise, but also by how well we handle ourselves and each other. This yardstick is increasingly applied in choosing who will be hired and who will not.' Clearly, the East Europeans interviewed by Anna failed this 'handling' test.

Vacancies in basic jobs – room cleaning, security and so forth – were filled locally rather than by overseas recruitment and largely through the use of employment agencies. BI used two agencies in 2006 to recruit casual staff. They were not directly employed by the hotel but worked under a variety of contracts, including zero-hours contracts (where staff are called in to cover busy periods but have no guaranteed hours), longer temporary contracts (from six months to two years), and, typically in the housekeeping division, contracts based on payment for the completion of a set number of tasks rather than on hours worked. Clearly, all these forms of 'flexible' employment are exploitative. Employees have little or no security of employment and few benefits such as holiday entitlement.

Recruiting through employment agencies brings substantial benefits to BI. The agencies not only screen potential employees but are their employers. As an informant from an employment agency commented:

> There is a financial and there's a political aspect to it, certain chains will outsource everything and the reason they do that is, it means they've got a low head count. Now it doesn't make any difference necessarily at the end because it obviously goes on their purchase ledger, the cost of using suppliers like us, but it's all about shareholder value at the end of the day, and of course a low head count, for some reason the city and the shareholders like that.... The less [*sic*] people you employ directly, the healthier it looks on your balance sheets.

She continued:

> In the event of a downturn, they will lose our staff, it will be our responsibility to re-employ, not the hotels' ... if they have permanent staff, they have contractual obligations. The hotels get a pure price, they don't pay any sickness

to this person, they do not pay the maternity, at any point they can say 'this person I do not want on the premises any more'. If she's our employee, we have to put her somewhere else.

This form of sub-contracted recruitment through local employment agencies had become a more significant element in BI's strategy over the last few years as a decision was made to out-source almost all the low-waged positions. This had an interesting effect on BI's employment policy, however, challenging the extent to which 'naming' by managers influences the recruitment process as BI became increasingly dependent on workers identified by employment agencies as suitable. This had an impact, as in 2005 a change of agency transformed the room attendants from Vietnamese men into Polish women. In both cases, however, stereotypical characteristics associated with nationality and gender were used by managers to construct and regulate the two labour forces. The 'submissiveness' of Vietnamese men, for example, was contrasted favourably with the 'aggressive' Polish women.

In the next section, the construction of difference in the labour market and its effects in segmenting the labour force are discussed. The different ways in which a gendered and racialized labour force is named, assembled, divided, maintained and resists its attribution, its work allocation and exploitation are explored.

Naming Workers for Specific Tasks

Like all hotels in large chains, administratively and organizationally BI is divided into a number of divisions – conference and events, food and beverage, front office and housekeeping. Each division had different recruitment practices, different embodied requirements, and consequently a different gender/ethnic/national division of labour. Not only the skill requirements but also whether the work was front or back office and whether or not it involved direct contact with guests affected recruitment strategies and the embodied attributes demanded in the performance of different sorts of tasks. There was an immediately visible distinction based, it seems, officially on nationality and associated language skills, but also on skin colour. Managerial positions were dominated by migrants from developed nations, largely Western Europe but also South Africa (the overall manager was German, the executive chief French, the executive housekeeper South African, the food and beverage manager Belgian). They were, without exception, white skinned. At lower levels in the organization, the proportion of staff from lower-income countries, including the Indian subcontinent and Eastern Europe, increased.

Front office

The front office team – which includes not only the front desk but the switchboard and the executive lounge – is among the most ethnically diverse in the hotel. The manager, Mark, a southern African, has a deliberate policy of constructing a multi-ethnic team. 'We are quite a big team ... we have in the front office I would say at least ten nationalities.' When Mark was appointed he took over a team consisting of mainly Indian, Pakistani and Chinese employees, but he has deliberately expanded his recruitment to include others, most recently Hungarian, Lithuanian and Polish workers. His team also includes individuals from Italy, the Netherlands, Korea, Sri Lanka, Germany, Jordan and Turkey. He feels this is important for the hotel guests 'because then they can see that we are a multinational team working in a multinational hotel', able to assist guests in their own language.

One of the most significant attributes for employment on the front desk is the ability to strike an immediate rapport with guests, 'managing' the exchange when guests are flustered and tired but remaining polite and welcoming. When Mark interviews candidates, his main criterion is 'are they able to strike up a conversation with me? They have never met me, so if they can do it with me, they should be able to do it with a guest that they are seeing for two or three minutes on a check-in or a checkout.' These skills are typically seen by managers as 'innate', as natural. As Mark said, 'if you haven't got it, then you can't learn it.'

Although Mark insisted that gender was not a criterion of employment, over 80 per cent of his staff are female. Here the classic associations between femininity and the presentation of self as available, willing and pleasant that have been identified so many times in different service sector occupations – from flight attendants (Hochschild 1983) to selling banking services (Halford, Savage and Witz 1997) – are key to managerial assessments of who is an appropriate front desk employee. Femininity typically is associated with servility, docility and deference, as well as with a pleasant demeanour, as the young men in the first half of this chapter found to their cost. Mark noted that he himself was atypical in the hotel industry as front office managers were usually women: affective intelligence and the management of emotions typically is regarded by hoteliers as a feminine trait. Mark's deputy – Sophia – is an Italian woman: combining a 'woman's touch' with a 'Latin temperament' (in Mark's words). But one of the advantages of masculinity, as Mark noted, is its association with authority and power. This parallels the 'assumed authority effect' documented by Simpson (2004) in her study of men in the female-dominated professions discussed in the last chapter. Hotel guests automatically assume that men are more senior than women and Mark sometimes finds that he has to intervene in disputes:

I reiterate exactly what my [female] colleague has said … we say exactly the same thing but they won't accept it from the assistant manager, which is just complete and utter nonsense, but there you are.

Bettina, on the other hand, a front desk supervisor from the Netherlands, believed that women are better at dealing with awkward guests:

Women get away with more on reception than men in terms of when you, for instance, can't give someone a certain type of room. You smile, fun, you make fun, you have a joke and it's fine. But if it's a man dealing with a man it gets a bit like … and then the supervisor needs to come out.

This interactive capacity is highly valued by hoteliers in the front desk employees and crucially depends on a deferential performance. As Waldinger and Lichter (2003: 50) noted:

To get the job done, one had to display the right face, and maintaining the appropriate front required a willingness to serve, or, at least, the ability to play the subordinate, good naturedly.

Rania, from Jordan, confirmed the significance of a performance that embodied good-natured service with a smile: 'It's very important, especially in reception … we're the first and the last person to see and the smile is very important' (and see Hall 1993a). And Janelle, a switchboard operator, suggested 'they have to hear the smile in your voice when you answer' – a task that she found so difficult sometimes that 'I just have to run in the back and scream, vent my emotions out of sight'.

The presentation of self involves more than a smiling face or voice, however. Appearance is also a key aspect of recruitment and the daily embodied performance required from the front office staff. As Rania noted, 'Appearance is not the be all and end all, but I think, yeah, we are on show, at the end of the day. We look at grooming, it's a standard within BI, you have to be groomed properly.' Several other employees in this area also used the term 'well groomed' to define the required look (as did employees in other divisions). But Mark also commented that the receptionists 'look elegant': a term seldom associated with a man. And Rania, with a series of gestures to her breasts and buttocks, emphasized that the typical embodied attributes of femininity were a significant part of her own success with guests, even though she had to police a line between flirtation and harassment. As Guerrier and Adib (2000) found, receptionists often have to cope with sexual innuendo from guests. Interestingly, Rania also mentioned that she had to deal with jealousy from her co-workers on the desk: 'The thing is that I have sometimes, it's girls, problem with my beauty.'

As well as physical attributes – slenderness, youth – assumptions about what is and is not an acceptable appearance on the front desk tend to discriminate between potential employees. Mark noted, for example, that nose rings and tattoos generally were unacceptable and that this practice discriminated against Asian women in the local labour market, if they reached an interview through an open application to the hotel. Employees of agencies are screened before they ever get to the hotel. Femininity (and masculinity) and appearance are, of course, raced and classed as well as gendered. In hospitality in general and in BI in particular, male and female employees from different class backgrounds and different nationalities are subject to different codes of judgement and so different daily experiences in their social interactions based on particular sets of assumptions and expectations among their co-workers, employers and the guests.

Here the dual nature of interpellation within a hotel is important, as managers make assumptions about what guests expect, and guests treat people unequally on the basis of a series of stereotypical assumptions. Thus the expectations of, for example, white customers of a Black, Asian or white female desk clerk may vary. Sophia, the assistant manager of the front office, commented:

> Sometimes guests get a little bit strange when it's perhaps, how can I say, international colleagues, so if you are not European, sometimes they can be a bit strange, especially Asians or if you are Black or something.

Thus Mark's stated desire to build an international team which had the effect of reducing the numbers of people of colour in his section may also reflect implicit racist assumptions. As Gabriel (1988: 4) noted, in the hospitality sector, service workers not only have to sell their labour (and bodies) to employers, but 'do so under the scrutiny of the customer who is paying to be served, obeyed and entertained'. In her work, Adkins (1995) noted the absence of Black women from front desk positions in a hotel and leisure park and suggested that this may be because they are not regarded as sufficiently sexually attractive for the predominantly white clientele. These patterns of discrimination based on stereotypical assumptions by managers and guests – the dual interpellation identified by Williams (2006) in her study of toy shops – are captured in Patricia Hill Collins' (2000) concept of a 'matrix of domination' in which hierarchies based on ethnicity and skin colour are evident. This matrix has the same effects as Butler's heterosexual matrix, which she defined as a 'grid of cultural intelligibility though which bodies, genders, and desires are naturalized' (Butler 1990: 151). In the same way, 'raced' and gendered bodies and cultural assumptions about their worth are also naturalized.

Conclusions: Warm Bodies and the International Division of Labour

As I have shown through the BI example, employment agencies increasingly are used to recruit workers in the hospitality industry. Bellman International, like many hotels, uses agencies to recruit workers to different positions, both for professional and less skilled vacancies. The use of agency and contract labour is expanding in the UK's 'flexible' labour market (Mangum, Mayall and Nelson 1985; Peck 1996; Storrie 2002; Forde and Slater 2005). In the public and the private sector, growing numbers of jobs and occupations are undertaken by workers either not directly employed by the organization in which they work or on short-term temporary contracts. In some cases, as in BI, the contract is for labouring bodies; in others, organizations contract the provision of a service (e.g. cleaning or security services by the National Health Service) and the workers involved in this provision are employees of the sub-contracting firm. In each case workers typically have little security and few employment-related benefits. Employment agencies are central in the assembly of this temporary workforce and yet the rise of the employment agency and of contract or contingent work is one of the least noticed and analysed phenomena in the construction of labour markets for service sector employment in advanced industrial economies' labour markets (although see Peck and Theodore 2001, 2002; Theodore and Peck 2002; Ward 2003, 2004; Peck, Theodore and Ward 2005; Coe, Johns and Ward 2006, 2007).

The use of temporary contract workers is particularly significant in economies where permanent workers have a high degree of employment protection but temporary workers do not (Booth, Francesconi and Frank 2002). Britain is one of the least regulated labour markets in the European Union, with few restrictions on the use of agency labour and fixed-term contracts, so it is not surprising that temporary work is significant. There was a rapid growth in contract employment throughout the 1980s in Britain associated with new forms of compulsory contracting-out of services such as catering and cleaning in the public sector, initiated by the Thatcher governments in the 1980s. Since then the numbers have remained stable (McOrmand 2004). In 2007 about 1.4 million British workers were agency workers of one kind or another. The net result of this growth of a precarious or contingent labour force, recruited and/or employed by employment agencies, is a deepening of the dual labour market discussed in chapter 2. The growing divergence in the terms and conditions of employment between the core of permanent workers and a periphery of temporary workers exacerbates the patterns of inequality between workers identified earlier.

Employment agencies operate at a range of spatial scales, assembling and providing labour to a range of firms and organizations, from multinational corporations to small, family owned businesses. The agencies themselves also vary in size and scope. At the top end of the labour market, staffing agencies operate across a global scale, connecting potential workers to vacancies in expanding and shortage sectors. Case studies from 'body shopping' in the computer industry, connecting Indian analysts with firms in Australia (Xiang 2006), to global agencies in the medical professions (Ball 2004; Bach 2007) bringing doctors from Asia, Africa and Australasia to work in the UK, have explored the growing interconnections constructed by employment agencies. Coe, Johns and Ward (2006, 2007) have begun to investigate the global reach of the staffing industry in producer service industries such as finance, telecommunications and IT, mapping its movement beyond core markets in North America and Western Europe into the emerging markets of Eastern Europe, Latin America and East Asia. The major players in this process of globalization are multinational firms owned by global capital. Adecco, for example, is the largest of these international agencies, registered in Switzerland, with global revenues exceeding US$22 billion and operating across 77 countries (Coe, Johns and Ward 2006: 5). At the bottom end of the labour market, Peck and others (Peck and Theodore 2001, 2002; Theodore and Peck 2002; Peck, Theodore and Ward 2005) have investigated the practices of 'hiring halls' in Chicago which provide 'warm bodies' for vacancies in local firms on a much more restricted spatial scale where 'most of the transactions are local ones – connecting local job seekers to local employers' (Peck and Theodore 2001: 23).

Bellman International is an example of an organization operating at an intermediate scale. Here, bottom-end vacancies are filled by transnational migrants, recruited not only locally within London but also by employment agencies operating in several European cities. For BI, small-scale national, rather than international, employment agencies, both 'here' and 'there', play a key part in assembling transnational labour forces. Even the lowest paid jobs at the bottom end of the hierarchy of status and reward are filled by transnational migrants. And even when transactions are local – between agencies, workers and potential employees in the same city – the putative employees may be extra-local. Thus, these employment agencies are reshaping larger-scale connections between nation-states, as well as playing an active part in the restructuring of racialized and gendered divisions of labour in British towns and cities. As I argued in chapter 3, the intersections between class, gender, ethnicity and nationality, as well as new connections across a differentiated global space-economy, are a key part of understanding the division of labour in the high-touch interactive service economy. Young, unskilled men in the UK labour market are not only at a disadvantage compared to women of their own age and older women re-entering the labour market, but also face increased competition from transnational workers, as do British–born members of minority communities.

9

Conclusions: Bodies in Place

None of us is outside or beyond geography
Edward Said, *Culture and Imperialism*, 1994

In this final chapter I want to reflect on the main arguments about continuity and change in service sector labour markets; restate my claim for the utility of a geographical analysis of work and employment, both in general and in embodied interactive work in particular; address a key methodological issue about the significance of case studies and ethnographies; and end with a question about embodiment in a world where virtual space and virtual identities seem to be becoming increasingly commonplace.

Continuity and Change in the Labour Market:
Grand Narratives, Crises and Dramas

I wrote this concluding chapter towards the end of 2008, when the terms 'economic transformation' and 'crisis' seemed to be at the centre of almost all debates – in the media, at the party political conferences held in the UK each autumn, in the final stages of the 2008 US election, in everyday conversations in shops, bars and cafés as rising prices and the credit crunch had a growing impact on the cost of living and as the British and US economies seemed to be sliding into recession. The over-hyped claims about the rise of a new knowledge economy began to sound hollow and, as Marx so memorably argued as capitalist social relations penetrated all areas of life, 'all that is solid melts into air'. Whether what is referred to as the 'money economy' (as distinct from the 'real economy') was ever solid is an interesting question, but it certainly seemed to be melting fast in 2008. The financial crisis on Wall Street and in the City of London, largely precipitated by reckless investments in the housing market and trading overly complex packages of debt between institutions, has seen the spectacular failure of large parts of the

`crisis' undermined
idea of change?`

banking industry in these two nations, as well as growing insecurity in Germany, France, Belgium, the Netherlands and Luxembourg and the collapse of the banking system in Iceland. The employees of failing banks and banks that have changed ownership were waiting anxiously to hear about their future employment prospects and as the stock market continued to fall, pension funds lost value and unemployment rates began to rise. A general mood of anxiety about economic prospects in general and labour market growth in particular started to pervade commentaries about waged work in many advanced industrial economies, as forecasts of rising unemployment became common.

It may seem perverse to argue at such a moment of crisis that the grand narratives of change that have dominated sociological and geographical analyses of labour market transformation over the last decades exaggerated the significance of the trends they purported to discern. In narratives of change, the focus tends to be on what is identified as the leading edge of the economy – that sector that has exhibited fastest growth, highest rates of profit, or overall contribution to the economy. Here, over the 1980s and 1990s, the financial services sector and other forms of knowledge-based growth – in science-based and high-tech industries for example, and in the arts, media and creative industries – met the criteria of rapid expansion and rising profits. These industries are also exciting and visible, especially in the capital, where they dominate the media based in the same cities. They are, as geographers Nigel Thrift, Andrew Leyshon and Peter Daniels (1987) once noted, 'sexy greedy' industries. If these were the industries that led the growth and now dominate the downturn in western economies that characterizes current economic discourse, then perhaps the epochal theorists are being proved right, albeit in a way they never forecasted. In retrospect the embedded arrogance of the debates about knowledge, thin air and masters of the universe invited disaster. There is something in the narratives of change themselves, and their focus on money and finance, of the same 'irrational exuberance' that marked the financial services sector itself throughout the 1990s. The term was coined in 1996 by Alan Greenspan when he was chairman of the US Federal Reserve Board and seemed even more appropriate in 2008 than it did twelve years earlier. There is a lot of hot air, as well as the thin air identified by Leadbeater (1999), in the way in which the financial services sector operates. It is, above all, an industry that relies on confident talk: once this evaporates, as it did at the end of 2008, the stock market falls, profits slump and talk of crisis becomes widespread. This is not, of course, an explanation for the current financial crisis, but rather an associated factor and crucial characteristic of the nature of the industry.

This focus on a particular sector of the economy and particular groups of workers characterized academic analyses of labour market change and the rise of a 'new' economy over the same 30-year period of financial sector expansion and economic growth in western economies. The new knowledge workers, it was argued, were freed from the ties of locality, roaming the space of flows, whether literally as embodied transnational workers or metaphorically in the virtual space constructed through new informational technologies. These workers were, according to Beck, Sennett, Castells and Bauman and other sociologists, increasingly freed from the traditional constraints of gender and class background, able to construct an individualized narrative and a flexible career portfolio in their search for highly paid work. Their idealized personification was as the 'masters of the universe' in Tom Wolfe's (1987) fictional account of Wall Street, although even here the local uneven geography of privilege and poverty cheek-by-jowl in New York City and a venture into the inner-city streets of Harlem brought disaster to the protagonist.

These masters of the universe and other knowledge-based workers are the ones who, according to Bauman, typically are able to fulfil their every desire for instant gratification in an expanding consumer economy where almost everything is for sale. But the theoretical and academic focus on these workers and their sexy/greedy lives seduced the analysts as much as the workers themselves. The less glamorous lives of other workers, tied to place and unable to satisfy every whim in the market, in large part escaped the lens of the epochal theorists. Nevertheless, in an economy marked by increasing inequality and polarization between the most affluent and the poorest workers, their lives were increasingly interconnected. For the most affluent, their instantaneous gratification of desire as almost everything under the sun became a commodity, as well as their ability to work across an expanding spatial canvas, depended on a growing army of workers to service their needs and desires, look after their dependents, clean their homes and offices and provide drink, food, entertainment and emotional solace, as well as run the transport and communication systems that permit geographical connections on a broader scale. As I argued in chapters 2 and 3, and showed in detail in the case study chapters, these embodied interactive workers have grown in number and in significance in the workforce, as well as become increasingly diverse in their social characteristics. Many men and growing numbers of women, from multiple national backgrounds, of different ages, abilities and skill levels, are now engaged in high-touch, interactive, emotional servicing employment, in a diversity of locations, from the home and the street to the specialized locations that characterized workplaces in the Fordist era. The expansion of this type of work also characterizes the years of irrational exuberance.

High-touch embodied service work is work that is demanding, often demeaning, spatially fixed, demands immediate response to the needs of the consumer, is typically technologically unsophisticated and characterized by low productivity. It depends on the embodied co-presence of the supplier and consumer of the service and a personal interaction between them. All these characteristics mean that these forms of servicing are costly to provide, it is difficult to increase the rate of profit, and necessitate low-paid forms of work for employees. As I have also shown, the associations between femininity, caring, emotions, waste, smells, desire and fleshy embodiment act in concert to code these forms of work as 'naturally' women's work, reinforcing low rates of pay, flexible forms of provision, low levels of organization and precarious attachments to the labour market. Thus they remain forms of work that are unattractive to many men, challenging their masculine identities in the labour market.

In exploring and explaining the rise of these forms of work as well as the workplace identities of workers in different types of embodied interactive work, I have provided a different narrative of change. Rather than escape from the bounds of the local and from categorical forms of identity that are the basis of social inequality – class, gender, nationality, citizenship rights and skin colour – interactive workers at the bottom end of the service economy are trapped in relations of social inequality in an economy increasingly characterized by extreme inequality in the labour market. The highest paid senior executives and managers of British and US firms, as well as the highest levels in government and the civil service, remain dominated by men. In a survey carried out by the *Guardian* newspaper and Reward Technology Forum (Teather 2008) of executive remuneration in firms listed in the FTSE100 (the most important firms listed on the Stock Exchange in the UK), there were 1,114 men and only 146 women (of whom 23 were executive directors) on the boards of directors in 2007. The differential between the highest paid executives and the average salary of their employees had widened substantially over the last decade. In 2007, in the retail industry, for example, the chief executive of Tesco earned an astonishing 526 times the average annual wage in his organization (£6,267,360 compared with £11,918) and at Sainsbury's 232 times (£3,001,000 compared with £12,960). These are examples of extreme inequality, influencing the widening gap between the top and bottom 10 per cent of the earnings distribution.

The expansion of low-paid employment is the obverse side of the growing significance of new forms of knowledge-based work. In Thrift, Leyshon and Daniels' (1987) terminology, this is less a 'sexy greedy' story than one of lives of servitude and desperation as more and more men and women from diverse backgrounds become enmeshed in the social relations of forms of waged work that reproduce and strengthen social inequalities of race, class and gender. These workers exist on low pay, often at or below the official minimum wage and yet, as the case studies have illustrated, they sometimes risk their safety in the performance of often demeaning services. Even so, as the voices of the workers in these jobs have shown, there is satisfaction to be found in the service of others, despite exploitation and insecurity, although this satisfaction cannot and should not compensate for poor pay. Despite the growth of these forms of work and the associated rise of women's labour market participation, I want to re-emphasize in this concluding chapter that the narrative is one of continuity as much as change. Rather than being a new form of work, embodied interactive service work has instead changed its location and shifted its designation from work to waged employment, so entering official definitions and accounts. Reproductive labour and servicing work has always been a significant part of economic growth and change and a large part of most women's lives. Despite its slow shift into the market and its growing significance as part of state-provided social services, these forms of work have a long history and a long association with femininity and lack of recognition. Nevertheless, its transfer into a form of waged employment is significant – a transfer that has accelerated significantly in the post-Fordist years in an economy based on the rise of consumerism and the associated growth of women's waged work.

Since the 1970s, for many women, their ability to earn an independent wage has increased significantly, as has their relative independence, changing

employment and gender.

The Spaces of Work: Place Matters

As I have argued, one of the most significant changes in the labour market has been the transfer of a good deal of interactive servicing from one place to another – from the home to the labour market, from the affective to the cash economy. At the same time, waged employment in the home has also grown in significance, challenging the associations between the domestic arena with social relations based on love, affection and co-sanguinity, as well as its definition as a private arena. The scale of class struggles has thus shifted down a notch from the specialized workplace into the home and become a more personal relationship than is perhaps typical between most employers and workers. In all types of work, however, but especially interactive service employment, the labour relation is a place-specific, localized one. As geographer Ray Hudson (2001: 122) has argued, 'labour is the most place-based of the factors of production'. As I noted in chapter 1, a fleshy worker has to turn up for work on a daily basis, fed, clothed and ready to put in the necessary effort, after a short or a long journey to work. Thus the available transport options, the cost of housing, the nature of the education system, the set of accessible job opportunities, as well as forms of social support for dependents in a locality, all affect an individual's propensity and opportunity to work and the ways in which employers assemble their labour force.

Paradoxically, the labour market is also an extra-local formation. Particular local labour markets are affected by national and supranational events. The financial crisis of 2007/8 provides a clear example. Poor investment strategies by bankers on Wall Street and in the City have reverberated across the UK. In September 2007 the problems of Northern Rock resulted in job losses in Newcastle, where the bank had its headquarters, and elsewhere. A year later, as HBOS (an organization that combined the Halifax and the Bank of Scotland) was taken over by Lloyds, workers in Halifax and Edinburgh were affected by uncertainty and rationalization. That same month, the Bradford and Bingley was partly nationalized and partly sold to a Spanish bank, Banco Santander, that had already swallowed up the Abbey National. Workers in Bradford and Bingley found themselves employed by a Spanish organization, stretching relationships across trans-European space. Labour, as well as capital, is also transnationally mobile. Filipina nannies working for international bankers, perhaps employed by Lehman Brothers in London – an investment bank that failed in 2008 – might have found themselves out of work, whereas nurses from India, the Caribbean or New Zealand perhaps had an increased workload as stressed bankers were hospitalized.

These patterns of migration within and between nation-states are affected by uneven spatial development, by variations in demand, by the structure of national and local labour markets and by the forms of regulation in place in national and transnational organizations and states that determine who is able to work where, as well as their eligibility for a range of rights and social benefits. The expansion of the European Union in 2004, for example, and later in 2007, restructured previous patterns of European migration. The right to seek waged work – available initially in only three of the 'old' EU 15 nations (Ireland, Sweden and the UK) – meant that the initial flows between the new EU 8 nations (Hungary, Czech Republic, Slovakia, Slovenia, Latvia, Lithuania, Estonia and Poland) were largely between these eleven countries. Indeed, the much larger than estimated number of European workers seeking work in the UK in the first two years after accession was a significant factor in placing restrictions on Bulgarian and Romanian workers when these nations joined the EU in January 2007. As illustrated in the case study at the end of chapter 8, these changing circumstances affected the origins of workers recruited as 'warm bodies' by employment agencies to work in a London hotel. In sex work, too, the changing patterns of international immigration are reflected on the streets of the capital and in smaller British towns and cities.

The insistence by geographers that work is an essentially geographical issue has been reflected in the changing emphases in the other social science disciplines. There is a great deal of work in economics, sociology and industrial relations that explores the nature and consequences of globalization, flows of capital, ownership of firms, worker migration, and trade union solidarity. But even so, the geographical imagination, in its insistence on interconnections across spatial scales, shows how local specificity – in the social relations in a boxing gym in Chicago or on the emergency ward in a London hospital, for example – provides a unique lens through which to explore the connections between workplace identities and the service economy. This specificity – explored here through case studies of particular occupations undertaken in a range of different workplaces in mainly British and some US cities – is composed of sets of social relations and flows of capital, people and ideas across space at multiple scales. Thus transnational movements, supranational and national legislation, regional and local variations in labour supply and demand, the uneven distribution of different types of work in different sizes and types of workplaces all affect who does what to whom, and where – the key question of the book first laid out in the preface. Through the three key concepts of place (the particularity or specificity I have just defined), space (the uneven geography or landscape of an economy and the connections across it) and scale (the different territorial spread of organizations and firms that constitutes both flows across space and the particularity of place), geographers are able to map the shape and flows of the uneven and unequal space economy and the social relations that define it. Space, unevenness and inequality are thus relational.

None of this is rocket science but, rather, a way of looking at a question that insists on always exploring the constitution of a labour force, the social relations of work and the divisions between employees in ways that emphasize their construction and maintenance through social relations that cross different spatial scales. Geography is a constitutive element in the construction of a labour force that both reflects and affects the way social processes work out over space.

Rather than simply serving as a stage upon which social life is played out or being merely a reflection of social relations, the construction of the economic landscape is particular ways is fundamental to how social systems function. (Herod, Rannie and McGrath-Champ 2007: 247)

Thus, to take an example from an earlier chapter, the relationships between the white residents and workers of a Chicago elder care facility or between the workers themselves cannot be understood without knowledge of the changing patterns of migration into and within the USA over the last century or more, the restructuring of the region that affected job opportunities as well as retirement incomes, the changing property market in the city that facilitated the transfer of hotels into care homes, and the changing ethnic composition of the city's population. Similarly in the boxing gym, the recent competition between Black and Hispanic boxers reflects the history of US migration policy, the transformation of the economies of Central American nations, the decline of manufacturing employment in Chicago, the extreme poverty of many inner-city residents and the particular associations between masculinity, embodied violence and minority men.

The grand narratives that have formed the story that I have been writing against in most of this book have tended to ignore the significance of the changing spatial division of labour, despite their general argument about growing interconnections across space in financial services, the legal profession and other high-status occupations, in part connected to the growth of multinational organizations with a labour force spread across more than one national territory. The interdependence between places and scales is largely neglected and the significance of the interconnections between the most local of social relations – whether in a gym, a hospital ward or in the trading rooms of a stock exchange – and larger-scale social processes is ignored. Although the specific case studies at the heart of this book have been of locally based workplace identities and practices, they illustrate connections across multiple spatial scales.

Two questions remain to be addressed:

1 Is it possible to generalize from the case studies and ethnographies of work and employment presented here?
2 How much does co-presence and embodiment still matter in service-dominated economies?

Generalizing from Case Studies

A key issue in studies of work and employment is the extent to which it is possible to generalize from ethnographies and case studies of particular workplaces. Is it possible to say anything more general from a single study or even a set of studies that rely on qualitative methodologies? In a review paper exploring this question, Edwards and Belanger (2008: 291) suggest that, although statistical generalization is not appropriate, it is possible to 'draw conclusions about how phenomena are connected in a "situated" manner'. They adopt an inclusive definition of ethnography as any study of a group of workers that uses observation or other ethnographic techniques. This definition encompasses almost all the studies I have drawn on here and extends well beyond the specific example of auto-ethnography. As I argued in chapter 6, auto-ethnographies raise particular issues about positionality for adherents of this approach.

Edwards and Belanger argued that the aim of producing comparative data should be built into individual ethnographies. In studies of employment, they suggest, a particular set of data might be collected across each individual case study, to include, for example, pay rates, absence levels, and rates of disciplinary action. Their particular interest is in workplace conflict and how it is managed in getting the work done. I would add to their suggestions an additional set of information about the social characteristics of a workforce and its hierarchical segmentation to allow arguments about embodied social characteristics and workplace performances to be compared. Edwards and Belanger also suggest asking respondents questions about similar types of work in different places, asking workers in areas of high unemployment, for example, how their position might differ in an area of low unemployment, to aid comparisons. Finally, they suggest more explicitly comparative work, suggesting that 'ethnography is ideally placed to trace through the effects of globalization on different forms of workplace regime' (Edwards and Belanger 2008: 310). Here, work comparing the division of labour in similar workplaces in different regions and different countries would be extremely revealing, showing the variation in assumptions about the legitimacy and suitability for differently situated bodies for different types of work.

While these arguments are interesting and their suggestions for ways to improve comparisons are useful, I want to argue, drawing on the claims in the previous section about the value of an explicitly geographical perspective, that the question about generalization is not always necessary and certainly not disabling. One of the key aims of an ethnography, a case study of particular types of employment or specific workplace interactions is to document the causes and consequences of the assembly of a labour force across interlocking spatial scales. Neither the local labour market itself, usually seen as coincident with a town or city, nor a particular workplace – whether a factory, a shopping mall or a house in a suburb – is conceptualized as a self-contained space, as a discrete entity, with an existence prior to the social relations between embodied workers and their clients. Thus, the place (at whatever spatial scale) is constituted through and by interconnections, just as previous sets of social relations affect the ways in which practices and places are defined and work out in a particular time-period. Place is not a box or a container, but rather a set of social relations. The purpose of case studies is not necessarily to search for representative examples, with the aim of generalizing from the findings, but is instead to understand complexity and particularity as the outcome of socio-spatial relations. The overall aim is to uncover the processes that lead to uneven development and inequality in labour markets, between both workers and regions that allows restless capitalism to use geographical difference to exploit workers and regions to increase profits (Harvey 1982). The uneven spatial landscape of capitalist development is not inevitable nor fixed, but changes over time, albeit taking particular shapes at different historical moments: the old Fordist landscape of manufacturing domination in the industrial heartlands of the west, is now being replaced with a new international division of labour in which newly emerging industrial powers are largely in Asia. As Harvey (1982: 416–17) argued, this pattern is 'actively produced rather than passively received as a concession to "nature" (the old argument about natural resource endowment) or "history"'. This active production shapes how social relations unfold over time and struggles against change or between workers or nations to attract 'mobile' capital investment.

It is this understanding of the mutual constitution of space and social relations (what the geographer Edward Soja (1989) termed the socio-spatial dialectic) that means that case studies of how workers interact, compete, struggle and unite in different spaces at different times are more than a single example at a particular moment in the organization of a labour process, the construction of a labour force and its segmentation, and the social relations between employers, workers and consumers. Even so, Edwards and Belanger's comments about ensuring comparisons are useful, reminding researchers to build into the research design both the collection of a set of basic data and common questions about the labour process and workers' interaction in order to understand labour market change and continuity. A research design that is directly comparative enables complicated questions to be investigated about the significance of place and the different ways in which embodied workers are evaluated and employed. Interesting examples that focus on the emergence of gendered divisions of labour include Helena Hirata's (1989) study of factories in Brazil and France, owned by an electronic multinational and making the same product – light bulbs – where she showed why a different gender division of labour emerged in each location; Milkman's (1987) study of the difference local labour market conditions made in the emergence of gendered divisions of labour in the Second World War in the USA; and Lee's (1997) work on how management choices and the regime of control in textile factories employing female labour in China and Hong Kong produced labour forces differentiated by age and status. As the auto-ethnographers insist, it is essential to connect case studies to theories of structural change.

Here, I have facilitated comparisons within and across the case studies through the establishment of a theoretical framework about the social construction and performance of difference. The combination of theoretical perspectives – a poststructural analysis of the body, sensitivity to the cultural meaning of class and gender divisions and an understanding of the political economy of globalization – provides a conceptual tool kit that helps explain the sort of new social and spatial divisions that are opening up in service economies and that are illustrated in detail through fine-grained qualitative research. Although case studies, ethnographies and interviews are the methods most commonly used in the work that I have introduced here, a range of other methods – the discursive analysis of novels and diaries, visual and content analysis of adverts, job specifications and photographs – are also appropriate in understanding contemporary representations of waged work. Indeed, the social sciences and the humanities are coming closer together in their mutual interest in the social relations in the workplace and there is scope for emerging work that builds on the use of multiple methods.

Meats

place in social arenas, their
sexuality become part of the interactive service that they are selling. However, in this part of the service sector as well as in other areas, including leisure pursuits such as computer gaming, but especially in the forms of social exchange becoming increasingly common among younger people, *escape* from the bonds of the body is an increasing possibility. A great deal of the work of investment bankers, for example, is done through virtual exchanges, manipulating bodies of information electronically on behalf of

invisible, often anonymous, clients, depending (perhaps erroneously, as it is turning out) on impersonal relations of trust, established by electronic communications. On social networking sites, anonymity, or rather masquerade, is also possible, as a virtual performance of any form of physical embodiment is within the reach of a key stroke.

The ability to escape from those fleshy characteristics of weight, gender, age and appearance that are such significant factors in the developing divisions of labour in the service economy have been most seriously discussed in the literature about cyberspace, although not in the specific case of social relationships in the workplace. In cyberspace – a virtual electronic reality existing independently from the material world – individuals are free to roam unhindered by a gendered, classed or raced body, removed from 'the immediacy of the immersion that affords no distance' (Grosz 2001: xv), in a new type of space unconstrained by the limits of corporeality. However, as Elizabeth Grosz has also noted, 'an awful lot of hype' (p. 74) surrounds debates about cyberspace and virtual reality. The term cyberspace itself seems to have been coined by the science fiction writer William Gibson in a short story *Burning Chrome* published in 1982 and then popularized in his well-known novel *Necromancer* (1984). Cultural fantasies about the impact of virtual reality are legion. Adherents 'believe there will be a choice not only of spaces, sites and environments but also of bodies, subjectivities and modes of interactions with others', whereas detractors 'fear and revile cyberspace's transformation of relations of sociality and community, physicality and corporeality, location and emplacement, sexuality, personal intimacy and shared work space – the loss of immediacy and physical presence' (Grosz 2001: 77). In cyberspace, love and bodily care, interactive, interpersonal and affective relations that structure the sorts of employment relations discussed here, are difficult to imagine and impossible to perform. Physical care for and interactions with others is, by definition, excluded from the spaces of virtual reality, although emotional exchanges – and sometimes emotional manipulation – are a characteristic of interaction in cyberspace. And the body is never entirely absent. 'Disembodied' explorers in virtual reality are still trapped in a corporeal frame, hunched over a keyboard, wrists aching from key strokes, eyes hurting from the glare of the screen, kept awake by caffeine-rich drinks.

For those who find the celebratory debates about cyberspace and virtual reality less than attractive, anxieties congeal around one particular term: the term 'meatspace' used by cybernauts and other IT afficianados to designate the material world and to distinguish it from disembodied virtual space. This, in my view, is a peculiarly offensive term that seems to have originated in the literatures of science fiction, and especially in cyberpunk novels. Meatspace encapsulates the real or physical world of bodies where interactions or transactions depend on physical co-presence as distinct from the disembodied interactions that take place in cyberspace. The term is derogatory, capturing the restrictiveness of the physical world, compared to the freedom of cyberspace. It seems to me too to allude to the desperation of 'meat markets' – sexual partner/date hunting events in pubs, clubs and other leisure spaces. Increasingly, these sites are part of cyberspace too, as Facebook and MySpace include dating sites. Like meatspace, cyberspace also comes with a set of social associations. Cyberpunk and cyberspace are for geeks, boys whose social confidence is too low to permit interaction beyond the keyboard. As William Gibson noted, moving through meatspace is a daunting prospect for true computer geeks.

Clearly, the term 'meat' is a dismissive reference to the human body, to the body itself as meat: as flesh, blood and bones. This is made clear by Frederick Pohl (1977) in his Gateway series (also a computer game), where he talks about meat intelligence and 'when I was meat' referring to individuals who have passed over to cyber intelligence. Interestingly, the startling image used on the cover of the first edition of Germaine Greer's *The Female Eunuch* (1970) was of a woman's torso hanging from a pole, for all the world like a piece of meat, reflecting/challenging the misogynistic view of women (then) widely evident. As I have argued in several chapters, the embodied associations between masculinity and femininity and different types of service sector occupations and jobs differentially advantage and disadvantage men and women. Dealing with the 'meatier' aspects of bodies – especially with the flesh, blood and dirt involved in physical aspects of caring – is associated with the female body and with women's 'natural' abilities to face bodily fluids and emissions with equanimity and to elide the differences between emotional labour, body work and dirty work in a single performance of feminine empathy. And this talent is, of course, located within the representations of women's own bodies as closer to nature, as fecund, unbounded and unruly (Merchant 1980; Butler 1990, 1993; Diprose 1994, 2002; Grosz 1994) and so as threatening in many of the rational and cerebral workplaces of the professions and in the high-tech economy. Bodily associations are drawn on the construct of women as out of place in these occupations, restricting many women to high-touch, co-present embodied occupations.

'The stickiness of material'

For women, escaping the bounds of the body seems particularly difficult, despite significant changes in women's lives over the last half century. Advances in healthcare and in contraception have reduced the physical costs of childbirth and most women now have fewer children than in previous generations. In the new millennium women are also members of the waged labour force in unprecedented numbers. And yet, as the preceding chapters have documented, femaleness still carries with it a penalty in job segregation and in a continuing gender gap in pay and in status. However, it is also increasingly clear that in a service-based economy in which forms of interactive work are numerically dominant, class, skin colour, ethnicity, age and weight, as well as gender, are all attributes of differentiation and discrimination in which potential workers are ranked in a hierarchy of desirability for different types of tasks and employment. Differently embodied men, as well as women, are also ranked in a hierarchy of desirability for particular types of jobs in service-based economies.

The servicing body, its emotions and affects, now matters in ways that differ from the labouring body of an industrial, manufacturing economy in which physical effort was more significant than personal presentation and social interactions. For workers in manufacturing industries, the end-user of the products being made remained unknown. Other than research on consumer preferences, the eventual purchasers of goods and their makers had no connections and certainly did not meet in person. The waged labourer producing the goods almost never interacted with their future owners and whether or not they might have personal opinions of each other was irrelevant. In contrast, in the personal interactions between providers and consumers that are an essential element of service sector occupations, a wide range of emotions fuels the exchange between embodied individuals. Disgust, contempt, shame, humiliation, anger, empathy, surprise, pleasure, enjoyment and excitement may singly or in some combination be part of the provision of a service that includes selling the body in different ways and these emotions may be felt by either or both workers and consumers. Many service workers find pleasure in their jobs, enjoying providing service, whether in a hotel, a hospital or a restaurant, a classroom or a high street shop, in a child's home or an elder care institution; but others do not. Selling sex is associated with fears of violence and anxiety about clients; selling burgers is often accompanied by a lack of respect from consumers; many aspects of care necessitate overcoming feelings of disgust and embarrassment. For most workers, emotional reactions to providing a service are usually mixed, as the pains and pleasures of work are often connected. Appearance, movement, gesture and behaviour and the 'messy' topics (Probyn 2004) of gender, sexuality and emotions are all part of the service encounter. Embodiment is now at the centre of economic life and how bodies connect (or do not) is a key issue in the service economy. As Grosz (2001: 83) noted, it is 'the stickiness of material that constitutes our entwinement with the real'. Interactive service work is both sticky and material when bodily presence is crucial and, as Foucault (1978) argued, the body is also the key site where strategies of control are felt and enforced, as well as the ultimate site of resistance to regulation.

The forms of work discussed in this book are not new, neither is their association with femininity; what has changed is both their location in the public arena and, because of the decline of manufacturing employment, the significance of these forms of low-waged work for men as well as for women. This emerging consumer service economy, which is one defined by low productivity gains and low wages, is producing an increasingly polarized economy and society. Class divisions are widening and new connections between class, gender and ethnicity are emerging. It is here in the associations between categorical inequalities and workplace identities that a claim to change lies. Despite the continuing patterns of gender inequality in the labour market, the relationships between gender and class have changed in the service economy now dominant in the UK, the USA and other western economies. Young men, for example, with low educational attainment find themselves particularly disadvantaged by economic change, as the traditional advantages of masculinity have deserted them in bottom-end jobs where deference is more highly valued than machismo or masculine bravado. Better educated, middle-class women are laying claim to higher-status jobs and so a new class division is opening between women. For poorly educated workers with few educational credentials, opportunities for achieving permanent employment in jobs that will provide a decent standard of living are diminishing. This is the other side of the optimistic accounts of a 'new' economy, in which high-tech, knowledge-based occupations are the drivers of change. These accounts now look increasingly hollow. For the high-status, high-tech workers not included in this book the privileges of private sector employment, especially the previously high rates of pay, now look less enticing than the more secure employment prospects in the public sector.

The workforce in the twenty-first century is also increasingly diverse as well as polarized, both in its composition and in the spaces in which labouring takes place. Not only are more women in waged employment, working in a range of situations and 'flexible' arrangements, but many workplaces now include people from many different nation-states and from different ethnic backgrounds, many, although not all of them, born outside the UK. Thus, connections between class, gender, nationality and ethnicity and skin colour are also being reconfigured in the UK. Growing numbers of white European migrant workers have entered the labour market since 2004, altering the relations between migration and identity established in the postwar decades, when migrants were more likely to come from the Caribbean or South Asia than from countries in East and Central Europe. While a report from a Select Committee on Economic Affairs in the House of Lords (2008) suggested that low-income workers in the UK, especially second generation migrants, made visible by their skin colour, may be losing out to more recent immigrants, the evidence is not yet clear, as the research base currently is limited. The new European migrants are 'transnational' workers (Westwood and Phizaklea 2000) in ways that people who came to Britain from the Caribbean from the late 1940s onwards and from India and East Africa from the 1960s generally were not (Brah 1996). These latter peoples have become settlers, despite the persistence of a myth of return (Bhachu 1985), visiting 'home' and, among the South Asian community, continuing connections with India, Pakistan and Bangladesh through marriage. The new Europeans may continue to be more mobile, moving between the UK and their home nations for employment as connections are easier and cheaper than for the 'new' Commonwealth migrants. Living within and between two (or more) states as European citizens with the right of entry to all EU member states may become an increasingly common pattern. Thus, the connections between class, gender, skin colour, status and national identity will become more fluid than in previous generations.

These patterns of economic migration between and within societies, as well as new patterns of class and gender relations, in large part a consequence of women's entry into service sector employment, are reshaping the British space economy. While almost all forms of interactive body work are local – depending on co-presence and face-to-face, if not always hands-on, connections between service providers and purchasers – new divisions of labour at multiple spatial scales mean an increasingly diverse labour force, assembled across a range of scales. New class, gender and ethnic social relations at work are thus being constructed, as workers in different class positions and of different national origins work within the previously 'private' spaces of the home, providing all sorts of intimate care for the body, as well as within more 'public' workspaces – in gyms and beauty parlours, hotels and fire stations. The growing influence of commoditized relationships within the home and of social relations based on emotions, love and care in workplaces blurs the boundaries between these spaces and challenges modernist conceptions of what is (or should be) private and what is public. The growing dominance of the high-touch and interactive servicing jobs thus has implications for both the social and spatial order of once-industrial economies, transforming relationships between men and women, between the social classes, between the indigenous populations and more recent immigrants, and between localities, regions and nations, placing gendered, classed and raced bodies at the centre of economic analysis.

There are implications too for the regulation of employment and for workers' strategies of organization and resistance. As I have shown through the case studies, many forms of bottom-end interactive work are precarious and casualized. Workers may not have formal contracts or guarantees of security of employment. Many are not members of trades unions. The huge variety of small workplaces in which interactive workers labour – from individual homes, shops and cafés to privatized care facilities – are far harder to organize than the large-scale factories of the Fordist era, while the increased diversity of the workforce makes solidarity harder to achieve. Only in the public sector – among the nurses, doctors and some care workers, as well as in local government and teaching – has trade union membership held up. Even here, terms and conditions of employment are deteriorating and for almost all workers fears about pension entitlements when they leave work are growing as occupational pension schemes become less generous as well as less common and more people are at the mercy of the stock market in saving for retirement. As employment becomes more central to more and more people's everyday lives and standards of living, it is also becoming more precarious, especially in the types of interactive employment described here. Ray Pahl suggested more than twenty years ago that work and working life is the key issue for social scientists. This claim is even stronger in the new millennium.

References

Acker, J. 1990 Hierarchies, jobs, bodies: a theory of gendered organizations, *Gender and Society* 4, 139–58.

Ackroyd, S. and Bolton, S. 1999 It is not Taylorism: mechanisms of work intensification in the provision of gynaecological services in an NHS hospital, *Work, Employment and Society* 13, 369–87.

Adkins, L. 1995 *Gendered Work: sexuality, family and labour market*, Open University Press, Buckingham.

Adkins, L. 2000 Objects of innovation: post-occupational reflexivity and retraditionalization of gender, in S. Ahmed, J. Kilby, C. Lury, M. McNeil and B. Skeggs (eds) *Transformations: thinking through feminisms*, Routledge, London.

Adkins, L. 2003 Reflexivity: freedom of habit of gender? *Theory, Culture and Society* 20, 21–42.

Adkins, L. 2004 Gender and the post-structural social. In B. Marshall and A. Witz (eds) *Engendering the Social*, Open University Press, Maidenhead.

Adkins, L. 2005 The new economy: property or personhood, *Theory Culture and Society* 22, 111–30.

Adkins, L. and Lury, C. 1996 The sexual, the cultural and the gendering of the labour market, in L. Adkins and V. Merchant (eds) *Sexualizing the Social: power and the organization of sexuality*, Macmillan, Basingstoke.

Adkins, L. and Lury, C. 1999 The labour of identity: performing identities, performing economies, *Economy and Society* 28, 598–614.

Adkins, L. and Lury, C. 2000 Making bodies, making people, making work, in L. McKie, L. Watson and N. Watson (eds) *Organizing Bodies: policy, institutions and work*, Macmillan, Basingstoke.

Adkins, L. and Skeggs, B. 2004 *Feminism after Bourdieu*, Blackwell, Oxford.

Aghatise, E. 2002 Trafficking for prostitution in Italy, paper presented at the expert group meeting on trafficking in women and girls, Glen Cove, NY, 18–22 November.

Agustin, L. 2000 A migrant world of services, *Social Politics* 1, 377–96.

Ahmed, S. 2004 Affective economies, *Social Text* 22, 114–39.

Alcock, P., Beatty, C., Fothergill, S., Macmillan, R. and Yeandle, S. 2003 *Work to Welfare: how men become detached from the labour market*, Cambridge University Press, Cambridge.

Alexander, J. and Mohanty, T. P. 1997 *Feminist Genealogies, Colonial Legacies, Democratic Futures*, Routledge, London.

Allen, J. 1992 Services and the UK space economy: regionalization and economic dislocation, *Transactions of the Institute of British Geographers* 17, 292–305.

Alvesson, M. 1998 Gender relations and identity at work: a case study of masculinities and femininities in an advertising agency, *Human Relations* 51, 113–26.

Amin, A. (ed.) 1994 *Post-Fordism: a reader*, Blackwell, Oxford.

Andall, J. 1992 Women migrant workers in Italy, *Women's Studies International Forum* 15, 4–48.

Anderson, B. 2000 *Doing the Dirty Work? The global politics of domestic labour*, Zed Books, London.

Anderson, B. 2001 Different roots in common ground: transnationalism and migrant domestic workers, *Journal of Ethnic and Migration Studies* 27, 673–83.

Anderson, B., Ruhs, M. and Spencer, S. 2006 *A Fair Deal? Migrant workers in low-wage occupations in the UK*, final report to the Joseph Rowntree Foundation (available from the authors at COMPAS, University of Oxford).

Anderson, L. 2006 Analytic autoethnography, *Journal of Contemporary Ethnography* 35, 373–95.

Andrews, E. 2007 Why half of us pay out to dodge those chores, *Daily Mail* 27 July, www.dailymail.co.uk.pages/live/article/news.html?in_article_id+470817, accessed 27.7.07.

Antipode 2006 Special issue on *Dirty work*, 38.

Arnot, M. 2004 Male working-class identities and social justice: a reconsideration of Paul Willis' 'Learning to Labour' in light of contemporary research, in N. Dolby and G. Dimitriadis (eds) *Learning to Labour in New Times*, Routledge, London.

Ashforth, B. E. and Humphrey, R. H. 1995 Emotion in the workplace: a reappraisal, *Human Relations* 48, 97–125.

Atkinson, J. 1985a The changing corporation, in D. Clutterbuck (ed.) *New Patterns of Work*, Gower, Aldershot.

Atkinson, J. 1985b Flexibility, uncertainty and manpower management, IMS Report No. 89, Institute of Manpower Studies, Brighton.

Atkinson, P., Coffey, A. and Delamont, S. 2003 *Key Themes in Qualitative Research: continuities and change*, Alta Mira Press, Walnut Creek, CA.

Augar, P. 2001 *The Death of Gentlemanly Capitalism*, Penguin, London.

Bach, S. 2007 Going global? The regulation of nurse migration in the UK, *British Journal of Industrial Relations* 45, 383–403.

Bailey, T. and Waldinger, R. 1991 Primary, secondary and enclave labor markets: a training systems approach, *American Sociological Review* 6, 432–45.

Bain, F. and Taylor, P. 2000 Entrapped by the 'electronic panopticon' – worker resistance in the call centre, *New Technology, Work and Employment* 15, 2–18.

Bakuan, A. and Stasiulis, D. 1997 *Not One of the Family: foreign domestic workers in Canada*, Toronto University Press, Toronto.

Ball, J. and Pike, G. 2004 *Stepping Stones: results from the RCN membership survey 2003*, www.rcn.org.uk/publications/pdf/membershipsurvey2003.pdf, Royal College of Nursing, London.

Ball, R. 2004 Divergent development, racialized rights: globalized labour markets and the trade of nurses – the case of the Philippines, *Women's Studies International Forum*, 27, 119–33.

Bartky, S. 1993 *Gender and Domination*, Routledge, London.

Bartoldus, E., Gillery, B. and Stuges, P. 1989 Stress and coping among home care-workers, *Health and Social Work* 14, 204–10.

Bates, I. 1993 A job which is 'right for me'? Social class, gender and individualization, in I. Bates and G. Riseborough (eds) *Youth and Inequality*, Open University Press, Buckingham.

Bates, I. 2005 Care work as body work, in D. Morgan, B. Branbdeth and E. Kvande (eds) *Gender, Bodies and Work*, Ashgate, Aldershot.

Bauder, H. 2001 Culture in the labour market: segmentation theory and perspectives of place, *Progress in Human Geography* 25, 37–52.

Bauder, H. 2002 Neighborhood effects and cultural exclusion, *Urban Studies* 39, 85–93.

Bauder, H. 2006 *Labour Movement: how migration regulates labour markets*, Oxford University Press, Oxford.

Bauman, Z. 1998 *Work, Consumerism and the New Poor*, Open University Press, Buckingham.

Bauman, Z. 2000 *Liquid Modernity*, Polity Press, Cambridge.

Beattie, G. 1996 *On the Ropes: boxing as a way of life*, Victor Gollancz, London.

Beck, U. 1992 *Risk Society: towards a new modernity*, Sage, London.

Beck, U. 1994 The reinvention of politics: towards a theory of reflexive modernization, in U. Beck, A. Giddens and S. Lash (eds) *Reflexive Modernization: politics, tradition and aesthetics in the modern social order*, Polity Press, Cambridge.

Beck, U. 2000 *The Brave New World of Work*, Polity Press, Cambridge.

Beck, U. and Beck-Gernsheim, E. 1996 *The Normal Chaos of Love*, Polity Press, Cambridge.

Beck, U., Giddens, A. and Lash, S. 1994 *Reflexive Modernization: politics, tradition and aesthetics in the modern social order*, Polity Press, Cambridge.

Bell, D. 1973 *The Coming of Post-Industrial Society*, Basic Books, New York.

Benaria, L. and Rodlan, M. 1987 *The Crossroads of Class and Gender: industrial homework, subcontracting and household dynamics in Mexico City*, University of Chicago Press, Chicago.

Bernhardt, A., Dresser, L. and Hatton, E. 2003 The coffee pot wars: unions and firm restructuring in the hotel industry, in E. Appelbaum, A. Bernhardt and R. Murnane (eds) *Low Wage America: how employers are reshaping opportunity in the workplace*, Russell Sage Foundation, New York.

Berthoud, R. 2007 *Work-Rich and Work-Poor*, Joseph Rowntree Foundation, Policy Press, Bristol.

Beynon, H. 1984 *Working for Ford*, Penguin, London.

Bhabha, H. 1994 *The Location of Culture*, Routledge, London.

Bhachu, P. 1985 *Twice Migrants*, Tavistock, London.

Biggs, D. 2005 Satisfaction levels amongst temporary agency workers, a review of the literature for the Recruitment and Employment Confederation (available from the author at University of Gloucestershire, Cheltenham).

Black, P. and Sharma, U. 2001 Men are real, women are 'made up': beauty therapy and the construction of femininity, *Sociological Review* 49, 100–16.

Bluestone, B. and Harrison, B. 1982 *The Deindustrialization of America: plant closings, community abandonment and the dismantling of basic industry*, Basic Books, New York.

Blunt, A. and Dowling, R. 2006 *Home*, Routledge, London.

Bolton, S. 2005 Women's work, dirty work: The gynaecology nurse as 'other', *Gender Work and Organization* 12, 169–86.

Bolton, S. and Boyd, C. 2003 Trolley dolly or skilled emotion manager? Moving on from Hochschild's *Managed Heart*, *Work, Employment and Society* 17, 289–308.

Bolton, S. and Houlihan, M. 2005 The (mis)representation of customer service, *Work, Employment and Society* 19, 685–703.

Bonnett, A. 2000 *White Identities: historical and international perspectives*, Prentice-Hall, London.

Booth, A., Delado, J. and Frank, J. 2002 Symposium on temporary work: introduction, *Economic Journal* 112, issue 480, 181–8.

Booth, A., Francesconi, M. and Frank, J. 2002 Temporary jobs: stepping stones or dead ends? *Economic Journal* 112, issue 480, 189–213.

Bordo, S. 1993 *Unbearable Weight: feminism, western culture and the body*, University of California Press, Berkeley.

Bourdieu, P. 1977 Remarques provisoires sur la perception sociale du corps, *Actes de la Recherche en sciences sociale* 14, 51–4.

Bourdieu, P. 1984 *Distinction: a social critique of the judgement of taste*, Routledge and Kegan Paul, London (originally published in 1979 as *La Distinction: critique social du jugement*, Editions du Minuet, Paris).

Bourdieu, P. 1990 *The Logic of Practice*, Polity Press, Cambridge.

Bourdieu, P. 1991 *Language and Symbolic Power*, Polity Press, Cambridge.

Bourdieu, P. 1999 *The Weight of the World: social suffering in contemporary society*, Stanford University Press, Stanford.

Bourdieu, P. 2000 *Pascalian Meditations*, Polity Press, Cambridge.

Bourdieu, P. 2001 *Masculine Domination*, Polity Press, Cambridge.

Bourgois, P. 1995 *In Search of Respect: selling crack in el barrio*, Cambridge University Press, Cambridge.

Bradley, H. 1989 *Men's Work, Women's Work*, Polity Press, Cambridge.

Bradley, H. 1993 Across the great divide, in C. Williams (ed.) *Doing Women's Work: men in non-traditional occupations*, Sage, London.

Brah, A. 1996 *Cartographies of Diaspora*, Routledge, London.

Brennan, D. 2004 *What's Love Got To Do With It? Transnational desires and sex tourism in the Dominican Republic*, Duke University Press, Durham, NC.

Brewis, J. and Linstead, S. 2000a 'The worst thing is the screwing' (1): consumption and the management of identity in sex work, *Gender, Work and Organization* 7, 84–97.

Brewis, J. and Linstead, S. 2000b 'The worst thing is the screwing' (2): Context and career in sex work, *Gender, Work and Organization* 7, 168–80.

Brewis, J. and Linstead, S. 2000c *Sex, Work and Sex Work: eroticizing organizations*, Routledge, London.

Brown, P. 1995 Cultural capital and social exclusion: some observations on recent trends in education, employment and the labour market, *Employment and Society* 9, 29–51.

Browne, J. and Minichiello, V. 1995 The social meanings behind male sex work: implications for sexual interactions, *British Journal of Sociology* 46, 223–48.

Browning, H. and Singelmann, J. 1975 *The Emergence of a Service Society*, National Technical Information Service, Springfield, MO.

Brush, L. 1999 Gender, work, who cares?! Production, reproduction, deindustrialization and business as usual, in M. M. Ferree, J. Lorber and B. Hess (eds) *Revisioning Gender*, Sage, London.

Bryman, A. 1999 The disneyization of society, *Sociological Review* 47, 25–47.

Bryman, A. 2004 *The disneyization of society*, Sage, London.

Bryson, J. and Daniels, D. 2007 *The Handbook of Service Industries*, Edward Elgar, Cheltenham.

Buchan, J., Jobanputra, R. and Gough, P. 2004 *London Calling? The international recruitment of health workers to the capital*, www.kingsfund.org.uk/pdf/international-recruitment/pdf, King's Fund, London.

Budd, L. and Whimster, S. (eds) 1992 *Global Finance and Urban Living*, Routledge, London.

Bunting, M. 2007 Sorry, Billie, but prostitution is not about champagne and silk negligees, *Guardian*, 8 October, p. 31.

Burawoy, M. 1979 *Manufacturing Consent*, University of Chicago Press, Chicago.

Butler, J. 1990 *Gender Trouble*, Routledge, London.

Butler, J. 1993 *Bodies that Matter*, Routledge, London.

Butler, J. 1997 *Excitable Speech: a politics of the performative*, Routledge, London.

Calvey, D. 2000 Getting on the door and staying there: a covert partipational study of bouncers, in G. L. Lee-Treweek and S. Linkogle (eds) *Danger in the Field: risk and ethics in social research*, Routledge, London.

Campbell, B. 1993 *Goliath: Britain's dangerous places*, Methuen, London.

Cameron, C., Mooney, A. and Moss, P. 2002 The child care workforce: current conditions and future directions, *Critical Social Policy* 22, 572–95.

Campkin, B. and Cox, R. (eds) 2007 *Dirt: new geographies of cleanliness and contamination*, Tauris, London.

Cannadine, D. 2000 *Class in Britain*, Penguin, London.

Carbado, D. (ed.) 1999 *Black Men on Race, Gender and Sexuality: a critical reader*, New York University Press, New York.

Carnoy, M. 2000 *Sustaining the New Economy: work, family and community in the information age*, Russell Sage Foundation and Harvard University Press, New York and Harvard, MA.

Carrier J. and Miller, D. (eds) 1998 *Virtualism: a new political economy*, Berg, Oxford.

Castells, M. 1977 *The Urban Question: a Marxist approach*, Edward Arnold, London.

Castells, M. 1996 *The Rise of the Network Society*, Blackwell, Oxford.

Castells, M. 2000 Materials for an exploratory theory of the network society, *British Journal of Sociology* 51, 5–22.

Castells, M. 2001 *The Internet Galaxy: reflections on the internet society*, Oxford University Press, Oxford.

Castles, S. and Miller, M. 2005 *The Age of Migration*, Palgrave Macmillan, Basingstoke.

Castree, N. 2007 Labour geography: a work in progress, *International Journal of Urban and Regional Research* 31, 853–62.

Castree, N., Coe, N., Ward, K. and Samers, M. 2004 *Spaces of Work: global capitalism and geographies of labour*, Sage, London.

Cavendish, R. 1982 *Women on the Line*, Routledge and Kegan Paul, London.

Chambliss, D. 1989 The mundanity of excellence: an ethnographic report on Olympic swimmers, *Sociological Theory* 7, 70–86.

Chapkis, W. 1997 *Live Sex Acts: women performing erotic labour*, Routledge, New York.

Chari, S. 2004 *Fraternal Capital: peasant-workers, self-made men and the globalization of provincial India*, Stanford University Press, Stanford.

Chari, S. and Gidwani, V. 2005 Introduction: grounds for a spatial ethnography of labor, *Ethnography* 6, 267–81.

Charlesworth, S. 2000 *A Phenomenology of Working Class Experience*, Cambridge University Press, Cambridge.

Chatterjee, P. 2001 *A Time for Tea: women, labour and post/colonial politics on an Indian plantation*, Durham, NC, Duke University Press.

Cheney, E. M. and Garcia Castro, M. 1991 *Muchachas No More: household workers in Latin America and the Caribbean*, Temple University Press, Philadelphia.

Choo, H. Y. 2006 Gendered modernity and ethnized citizenship: North Korean settlers in contemporary South Korea, *Gender and Society* 20, 576–604.

Christie, H., Munro, M. and Rettig, H. 2001 Making ends meet: student incomes and debt, *Studies in Higher Education* 26, 363–83.

Clarke, J. and Salt, J. 2003 Work permits and foreign labour in the UK: a statistical review, *Labour Market Trends*, November, 563–74.

Coates, S. 1999 An ethnography of boxing: the role of the coach, *Sociology of Sport Online* 2, 1, pp. 1–4, www.physed.otago.ac.nz/sosol/v2il/v2ils1.htm.

Cockington, J. and Marlin, L. 1995 *Sex Inc.: true tales from the Australian sex industry*, Ironbark Pan Macmillan, Sydney.

Coe, N., Johns, J. and Ward, K. 2006 Mapping the globalization of the temporary staffing industry, *The globalization of the temporary staffing industry working paper series* no. 2, October, University of Manchester School of Geography, www.sed.manchester.ac.uk/geography/research/tempingindustry/download/wp 2.pdf.

Coe, N., Johns, J. and Ward, K. 2007 Managed flexibility: labour regulation, corporate strategies and market dynamics in the Swedish temporary staffing industry, *The globalization of the temporary staffing industry working paper series* no. 3, February, University of Manchester School of Geography, www.sed.manchester.ac.uk/geography/research/tempingindustry/download/wp3.pdf.

Cohen, R. 2008 Body work, employment relations and the labour process, paper given at seminar 1: The social significance of body work, ESRC Seminar, Body work: critical themes, future agendas, Univesity of Warwick, 18–19 January, www.go.warwick.ac.uk/bodywork.

Cohen, S. 1973 *Folk Devils and Moral Panics*, Paladin, St Albans.

Cohen, S. and Taylor, M. 1976 *Escape Attempts: the theory and practice of resistance to everyday life*, Allen Lane, London.

Collins, P. H. 2000 *Black Feminist Thought*, Routledge, London.

Collinson, M. and Collinson, D. 1996 'It's only Dick': the sexual harassment of women managers in insurance sales, *Work, Employment and Society* 10, 29–56.

Collison, D. and Hearn, J. 1994 Naming men as men: implications for work, organization and management, *Gender, Work and Organization* 1, 2–22.

Comaroff, J. and Comaroff, J. L. 2000 Millennial capitalism: first thoughts on a second coming, *Public Culture* 12, 291–343.

Connell, R. W. 1995 *Masculinities*, Polity Press, Cambridge.

Connell, R. W. 2000 *The Men and the Boys*, Polity Press, Cambridge.

Connell, R. W. 2001 The social organization of masculinity, in S. Whitehead and F. Barrett (eds) *The Masculinities Reader*, Polity Press, Cambridge.

Coser, L. 1974 *Greedy Institutions: patterns of undivided commitment*, Free Press, New York.

Courtney, C. and Thompson, P. (eds) 1996 *City Lives: the changing voices of British finance*, Methuen, London.

Coutts, K., Glyn, A. and Rowthorn, B. 2007 Structural change under New Labour, *Cambridge Journal of Economics* 31, 845–61.

Cowan, R. S. 1983 *More Work for Mother: the ironies of domestic technology form the open hearth to the microwave*, Basic Books New York.

Cox, R. 2006 *The Servant Problem: domestic employment in a global economy*, Tauris, London.

Cox, R. and Narula, R. 2003 Playing happy families: rules and relationships in au pair employing households in London, England, *Gender, Place and Culture* 10, 333–44.

Coy, M., Hovarth, M. and Kelly, L. 2007 'It's just like going to the supermarket': men buying sex in East London, Child and Woman Abuse Studies Unit, London Metropolitan University, London.

Coyle, D. 1997 *The Weightless World*, Capstone, Oxford.

Crang, P. 1994 It's showtime: on the workplace geographies of display in a restaurant in Southeast England, *Society and Space: Environment and Planning D* 12, 675–704.

Cressey, P. 1932 *The Taxi-Dance Hall: a sociological study in commercialized recreation and city life*, University of Chicago Press, Chicago.

Crompton, R. 1999 *Restructuring Gender Relations and Employment*, Oxford University Press, Oxford.

Crompton, R. and Sanderson, K. 1990 *Gendered Jobs and Social Change*, Unwin Hyman, London.

Crossley, N. 2006 In the gym: motives, meaning and moral careers, *Body and Society* 12, 23–50.

Cully, M., Woodland, S., O'Reilly, A. and Dix, G. 1999 *Britain at Work*, Routledge, London.

Daiski, I. and Richards, E. 2007 Professionals on the sidelines: the working lives of bedside nurses and elementary core French teachers, *Gender, Work and Organization* 14, 210–31.

Daly, M. and Rake, K. 2003 *Gender and the Welfare State: care, work and welfare in Europe and the USA*, Polity Press, Cambridge.

Dandeker, C. and Mason, D. 2001 The British armed services and the participation of minority ethnic communities, *Sociological Review* 49, 219–45.

Davidoff, L. 1988 Class and gender in Victorian England, in J. Newton, M. Ryan and C. Wolkowitz (eds) *Sex and Class in Women's History: essays from feminist studies*, Routledge and Kegan Paul, London.

Davidson, J., Smith, M. and Bondi, L. (eds) 2005 *Emotional Geographies*, Ashgate, Aldershot.

Davis, K. 1995 *Reshaping the Female Body: the dilemma of cosmetic surgery*, Routledge, London.

Davis, K. 2002 A dubious equality: men, women and cosmetic surgery, *Body and Society* 8, 49–65.

Davis, M. 1990 *City of Quartz: excavating the future in Los Angeles*, Verso, London.

Day, S. 2007 *On the Game: women and sex work*, Pluto Press, London.

Deery, S., Iverson, R. and Walsh, J. 2002 Work relations in telephone call centres: understanding emotional exchanges and employee withdrawal, *Journal of Management Studies* 39, 471–96.

Delamont, S. 2000 The anomalous beasts: hooligans and the sociology of education, *Sociology* 34, 95–111.

De Lauretis, T. 1987 The technology of gender, in T. de Lauretis (ed.) *Technologies of Gender: essays on theory, film and fiction*, University of Indiana Press, Bloomington.

Dennis, N., Henriques, F. and Slaughter, C. 1956 *Coal is Our Life*, Eyre and Spottiswoode, London.

Denzin, N. and Lincoln, Y. (eds) 2000 *Handbook of qualitative research*, Sage, Thousand Oaks, CA.

Department of Trade and Industry (DTI) 2001 *UK Competitiveness Indicators*, 2nd edn., Department of Trade and Industry, London.

Desmond, M. 2007 Becoming a firefighter, *Ethnography* 7, 387–421.

Devine, F. 2004 *Class Practices*, Cambridge University Press, Cambridge.

Devine, F., Savage, M., Scott, J. and Crompton, R. (eds) 2005 *Rethinking Class: culture, identities, lifestyle*, Palgrave Macmillan, Basingstoke.

Diamond, T. 1992 *Making Gray Gold: narratives of nursing home care*, University of Chicago Press, Chicago.

Dicken, P. 2007 *Global Shift: reshaping the global economic map in the 21st century* Sage, London.

Dickens, R. and Manning, A. 2002 Has the national minimum wage reduced UK wage inequality? Centre for Economic Performance Discussion Paper, London School of Economics, London.

Diprose, R. 1994 *The Bodies of Women: ethics, embodiment and sexual difference*, Routledge, London.

Diprose, R. 2002 *Corporeal Generosity*, State University of New York Press, Albany.

Di Stefano, C. 1990 Dilemmas of difference: feminism, modernity and postmodernism, in L. Nicholson (ed.) *Feminism/Postmodernism*, Routledge, London.

Donath, S. 2000 The other economy: a suggestion for a distinctively feminist economics, *Feminist Economics* 6, 116–23.

Donzelot, J. 1980 *The Policing of Families*, Hutchinson, London.

Doogan, K. 2001 Insecurity and long-term employment, *Work, Employment and Society* 15, 419–41.

Dorling, D., Rigby, J., Wheeler, B., Ballas, D., Thomas, B., Fahmy, E., Gordon, D. and Lupton, R. 2007 *Poverty, Place and Wealth in Britain, 1968 to 2005*, Policy Press, Bristol and Joseph Rowntree Foundation, York.

Douglas, M. 1966 *Purity and Danger: an analysis of concepts of pollution and taboo*, Routledge and Kegan Paul, London.

Dowling, R. 1999 Classing the body, *Environment and Planning D: Society and Space* 17, 511–14.

Du Gay, P. 1996 *Consumption and Identity at Work*, Sage, London.

Du Gay, P. 2004 The tyranny of the epochal and work identity, in T. Jensen and A. Westenholz (eds) *Identity in the Age of the New Economy*, Edward Elgar, Cheltenham.

Du Gay, P. and Pryke, M. 2002 *Cultural Economy: cultural economics and commercial life*, Sage, London.

Dworkin, A. 1974 *Woman-hating*, Dutton, New York.

Dyer, C. 2008 New sexual harassment law to protect staff from customers, *Guardian*, 31 March, p. 14.

Dyer, R. 1998 *White*, Routledge, London.

Dyer, S., McDowell, L. and Batnizky, A. 2008 Emotional work/body work: the caring labours of migrants in the UK's National Health Service, *Geoforum* 39, 2030–8.

Edwards, P. and Belanger, J. 2008 Generalizing from workplace ethnographies: from induction to theory, *Journal of Contemporary Ethnography* 37, 291–314.

Ehrenreich, B. 1984 Life without father: reconsidering socialist feminist theory, *Socialist Review* 73, 48–57.

Ehrenreich, B. 2001 *Nickel and Dimed: on (not) getting by in America*, Metropolitan Books, New York.

Ehrenreich, B. and Hochschild, A. (eds) 2003 *Global Woman: nannies, maids and sex workers in the new economy*, Granta, London.

Elliott, L. and Atkinson, D. 1998 *The Age of Insecurity*, Verso, London.

England, K. (ed.) 1996 *Who Will Mind the Baby? Geographies of childcare and working mothers*, Routledge, London.

England, P. 2005 Emerging theories of care work, *Annual Review of Sociology* 31, 381–99.

England, P., Budig, M. and Folbre, N. 2002 Wages of virtue: the relative pay of care work, *Social Problems* 49, 455–73.

England, P. and Folbre, N. 1999 The cost of caring, *Annals of the American Academy of Political and Social Science* 561, 39–51.

Erickson, R. J. and Ritter, C. 2001 Emotional labor, burnout, and inauthenticity: does gender matter? *Social Psychology Quarterly* 64, 146–63.

Erikson, B., Albanese, P. and Drakulic, S. 2000 Gender on a jagged edge: the security industry, its clients and the reproduction and revision of gender, *Work and Occupations* 27, 294–318.

Erikson, K. 2004 To invest or detach? Coping strategies and workplace culture in service work, *Symbolic Interaction* 27, 549–72.

Esping Andersen, G. 1990 *The Three Worlds of Welfare Capitalism*, Polity Press, Cambridge.

Esping Andersen, G. 1999 *Social Foundations of Welfare Capitalism*, Oxford University Press, Oxford.

Etzioni, A. 1969 *The Semi-Professions and their Organization*, Free Press, New York.

European Commission 1996 Communication from the Commission to the Council and the European Parliament for further actions on trafficking in women for the purposes of sexual exploitation, COM (96) 567, Brussels.

European Commission 2001 *Employment and social policies: a framework for investing in quality.* Communication from the Commission to the Council, the European Parliament, the Economic and Social Committee and the Committee of the Regions, COM (2001) 313 final, Brussels.

European Commission 2002 *Employment in Europe 2002*, European Commission, Directorate-General for Employment and Social Affairs, Brussels.

European Trade Union Federation 2007 *Temporary Agency Workers in the European Union*, www.etuc.org/a/50, Brussels.

Evans, J. 1997 Men in nursing: exploring the male nurse experience, *Nursing Enquiry* 4, 142–5.

Featherstone, M., Hepworth, M. and Turner, B. S. (eds) 1991 *The Body, Social Process and Cultural Theory*, Sage, London.

Felstead, A., Gallie, D. and Green, F. 2002 *Work Skills in Britain 1986–2001*, DfES Publications, Nottingham.

Fernandez Kelly, P. 1994 Towanda's triumph: social and cultural capital in the transition to adulthood in the urban ghetto, *International Journal of Urban and Regional Research* 18, 88–111.

Ferree M. M., Lorber, J. and Hess, B. B. (eds) (1999) *Revisioning Gender*, Sage, London.

Fine, G. 1998 Justifying work: occupational rhetoric in restaurant kitchens, *Administrative Science Quarterly* 41, 90–112.

Fine, M. and Weiss, L. 1998 *The Unknown City: the lives of poor and working class young adults*, Beacon Press, Boston.

Finney, A. 2004 *Violence in the Night-Time Economy: key findings from the research*, Research, Development and Statistics Directorate, Home Office, London.

Fisher, M. 2007 Enterprising women: remaking gendered networks on Wall Street in the new economy, in D. Perrons, C. Fagan, L. McDowell, K. Ray and K. Ward (eds) *Gender Divisions and Working Time in the New Economy*, Edward Elgar, London.

Flax, J. 1990 *Thinking Fragments: psychoanalysis, feminism and postmodernism in the contemporary west*, University of California Press, Berkeley.

Folbre, N. and Nelson, J. A. 2000 For love or money – or both? *Journal of Economic Perspectives* 14, 123–40.

Fonow, M. and Cook, J. 2005 Feminist methodology: new applications in the academy and public policy, *Signs: Journal of Women in Culture and Society* 30, 2009–15.

Forde, C. and Slater, G. 2005 Agency working in Britain: character, consequences and regulation, *British Journal of Industrial Relations* 43, 249–72.

Forseth, U. 2005 Gender matters? Exploring how gender is negotiated in service encounters, *Gender, Work and Organization* 12, 440–59.

Fothergill, S. and Wilson, I. 2007 A million off incapacity benefit: how achievable is Labour's target? *Cambridge Journal of Economics* 31, 1007–24.

Foucault, M. 1978 *The History of Sexuality Vol. 1: An Introduction*, Allen Lane, London.

Foucault, M. 1986 *The History of Sexuality Vol. 2: The Use of Pleasure*, Viking, London.

Frankenberg, R. 1993 *White Women, Race Matters: the social construction of whiteness*, Routledge, London.

Frankenberg, R. (ed.) 1997 *Displacing Whiteness: essays in social and cultural criticism*, Duke University Press, Durham, NC.

Freeman, C. 2000 *High Tech and High Heels in the Global Economy*, Duke University Press, Durham, NC.

Frenkel, S., Tam M., Korczynski, M. and Shile, K. 1998 Beyond bureaucracy? Work organization in call centres, *International Journal of Human Resource Management* 9, 957–79.

Frost, J. 2005 Theorizing the young woman in the body, *Body and Society* 11, 63–85.

Fryer, J. 1984 *Staying Power*, Pluto Press, London.

Fudge, J. and Owens, R. (eds) 2006 *Precarious Work, Women and the New Economy*, Hart, Oxford.

Furman, F. 1997 *Facing the Mirror: older women and the beauty shop culture*, Routledge, London.

Gabriel, Y. 1988 *Working Lives in Catering*, Routledge, London.

Gabriel, Y. 2004 The glass cage, flexible work, fragmented consumption, fragile selves, in J. Alexander, G. Marx and C. Williams (eds) *Self, Social Structure and Beliefs*, University of California Press, Berkeley.

Gardiner, J. 1997 *Gender, Care and Economics*, Macmillan, Basingstoke.

Gartner, A. and Riessman, F. 1974 *The Service Society and the Consumer Vanguard*, Harper and Row, New York.

Geertz, C. 1988 *Works and Lives: the anthropologist as author*, Stanford University Press, Palo Alto.

Gershuny, J. 1978 *After Industrial Society? The emerging self-service economy*, Macmillan, London.

Gerth, H. and Wright Mills, C. 1964 *Character and Social Structure*, Harcourt Brace Jovanovich, New York.

Gherardi, S. 1995 *Gender, Symbolism and Organizational Cultures*, Sage, London.

Giddens, A. 1991 *Modernity and Self-Identity: self and society in the late modern age*, Polity Press, Cambridge.

Giddens, A. 1992 *The Transformation of Intimacy: sexuality, love and eroticism in modern societies*, Polity Press, Cambridge.

Gill, R., Henwood, K. and Mclean, C. 2005 Body projects and the regulation of normative masculinity, *Body and Society* 11, 37–62.

Gimlin, D. 1996 Pamela's place: power and negotiation in the hair salon, *Gender and Society* 10, 505–26.

Gimlin, D. 2002 *Body Work: beauty and self-image in American culture*, University of California Press, Berkeley.

Glenn, E. N. 1992 From servitude to service work – historical continuities in the racial division of paid reproductive labour, *Signs: A Journal of Women and Culture* 18, 1–43.

Glenn, E. N. 2001 Gender, race and the organization of reproductive labour, in R. Baldoz, C. Koeber and P. Kraft (eds) *The Critical Study of Work: labor, technology and global production*, Temple University Press, Philadelphia.

Glucksmann, M. (aka Ruth Cavendish) 2009 *Women on the Line*, Routledge, London.

Goffman, E. 1959 *The Presentation of Self in Everyday Life*, Penguin, London.

Goleman, D. 1998 *Working with Emotional Intelligence*, Bloomsbury, London.

Goos, M. and Manning, A. 2003 McJobs and Macjobs: the growing polarization of jobs in the UK, in R. Dickens, P. Gregg and J. Wadsworth (eds) *The Labour Market Under New Labour*, Palgrave Macmillan, Basingstoke.

Goos, M. and Manning, A. 2007 Lousy and lovely jobs: the rising polarization of work in Britain, *Oxford Review of Economic Policy* 7, 49–62.

Gorz, A. 1982 *Farewell to the Working Class: an essay on post-industrial socialism*, Pluto Press, London.

Green, F. 2006 *Demanding Work: the paradox of job quality in the affluent economy*, Princeton University Press, Princeton.

Greer, G. 1970 *The Female Eunuch*, Flamingo, London.

Gregory, D. and Urry, J. (eds) 1986 *Social Relations and Spatial Structures*, Macmillan, London.

Gregson, N. and Lowe, M. 1994 *Servicing the Middle Class: class, gender and waged domestic labour in contemporary Britain*, Routledge, London.

Grimshaw, D. and Rubery, J. 2007 Undervaluing women's work, Working Paper Series No. 53, Equal Opportunities Commission, Manchester.

Grosz, E. 1994 *Volatile Bodies: toward a corporeal feminism*, Indiana University Press, Bloomington.

Grosz, E. 2001 *Architecture from the Outside: essays on real and virtual space*, MIT Press, Cambridge, MA.

Gubrium, J. 1975 *Living and Dying at Murray Manor*, St Martin's Press, New York.

Gubrium, J. F. 2007 Urban ethnography of the 1920s working girl, *Gender, Work and Organization* 14, 233–58.

Guerrier, Y. and Adib, A. 2000 'No, we don't provide that service': the harassment of hotel employees by customers, *Work, Employment and Society* 14, 689–705.

Gulcar, L. and Ilkkaracan, P. 2002 The 'Natasha' experience: migrant sex workers from the former Soviet Union and Eastern Europe in Turkey, *Women's Studies International Forum* 25, 411–21.

Haenfler, R. 2004 Manhood in contradiction: the two faces of Straight Edge, *Men and Masculinities* 7, 77–99.

Halbert, C. 1997 Tough enough and woman enough: stereotypes, discrimination and impression management among women professional boxers, *Journal of Sport and Social Issues* 21, 7–36.

Hale, A. and Wills, J. 2005 (eds) *Threads of Labour*, Blackwell, Oxford.

Halford, S. 2003 Gender and organizational restructuring in the National Health Service: performance, identity and politics, *Antipode* 35, 286–308.

Halford, S. and Leonard, P. 2006 Place, space and time: contextualizing workplace subjectivities, *Organization Studies* 27, 657–76.

Halford, S., Savage, M. and Witz, A. 1997 *Gender, Careers and Organizations: current developments in banking, nursing and local government*, Macmillan, Basingstoke.

Hall, E. 1993a Smiling, deterring and flirting: doing gender by giving 'good service', *Work and Occupations* 20: 452–71.

Hall, E. 1993b Waitering/waitressing: engendering the work of table servers, *Gender and Society* 17, 329–46.

Hall, S. 1990 Cultural identity and diaspora, in J. Rutherford (ed.) *Identity: community, culture, difference*, Lawrence and Wishart, London.

Hall, S. 1995 New cultures for old. In D. Massey and P. Jess (eds) *A Place in the World? Places, cultures and globalization*, Oxford University Press, New York.

Hamnett, C. 2003 *Unequal City: London in the global arena*, Routledge, London.

Hampson, I. and Junor, A. 2005 Invisible work, invisible skills: interactive customer service as articulation work, *New Technology Work and Employment* 20, 166–81.

Handy, C. 1989 *The Age of Unreason*, Arrow, London.

Hanson, S. and Pratt, G. 1995 *Gender, Work and Space*, Routledge, London.

Harding, J. 2000 *The Uninvited: refugees at the rich man's gate*, Profile Books, London.

Harvey, D. 1982 *The Limits to Capital*, Blackwell, Oxford.

Harvey, D. 1989 *The Condition of Post-Modernity*, Blackwell, Oxford.

Harvey, D. 2001 *Spaces of Capital: towards a critical geography*, Edinburgh University Press, Edinburgh.

Hawkes, G. 1996 *A Sociology of Sex and Sexuality*, Open University Press, Buckingham.

Haylett, C. 2001 Illegitimate subjects? Abject whites, neoliberal modernization and middle-class multiculturalism, *Environment and Planning D: Society and Space* 19, 351–79.

Hearn, J. 1982 Notes on patriarchy, professionalization and the semi-professions, *Sociology* 16, 184–202.

Heath, A. and Cheung, S. 2008 *Unequal Chances: ethnic minorities in western labour markets*, Proceedings of the British Academy, Oxford University Press, Oxford.

Heikes, J. 1992 When men are in the minority: the case of men in nursing, *Sociological Quarterly* 32, 389–401.

Herod, A. 2000 Workers and workplaces in the neo-liberal economy, *Environment and Planning A* 32, 1781–90.

Herod, A. 2001 *Labor Geographies*, Guilford Press, New York.

Herod, A., Rannie, A. and McGrath-Champ, S. 2007 Working space: why incorporating the geographical is central to theorizing work and employment practices, *Work, Employment and Society* 21, 247–64.

Hinsliff, G. 2007 Trafficked sex workers win right to stay, *Observer*, 21 January.

Hirata, H. 1989 Production relocation: an electronics multinational in France and Brazil, in D. Elson and R. Pearson (eds) *Women's Employment and Multinationals in Europe*, Macmillan, London.

Hobbs, D., Hadfield, P., Lister, S. and Winlow, S. 2002 Door lore: the art and economics of intimidation, *British Journal of Criminology* 42, 352–70.

Hobbs, D., Lister, S., Winlow, S. and Hadfield, P. 2002 *Night Moves: bouncers, violence and governance in the night time economy*, Oxford University Press, Oxford.

Hobbs, D., O'Brien, K. and Westmarland, L. 2007 Connecting the gendered door: women, violence and doorwork, *British Journal of Sociology* 58, 21–38.

Hochschild, A. 1983 *The Managed Heart: commercialization of human feeling*, University of California Press, Berkeley.

Hochschild, A. 1997 *The Time Bind: when work becomes home and home becomes work*, Metropolitan Books, New York.

Hochschild, A. 2003 *The Commercialization of Intimate Life: notes from home and work*, University of California Press, Berkeley.

Hoigard, C. and Finstad, L. 1992 *Backstreets: prostitution, money and love*, Polity Press, Cambridge.

Hondagneu-Sotelo, P. 2001 *Domestica: immigrant workers cleaning and caring in the shadows of affluence*, University of California Press, Berkeley (new edition with a new preface published in 2007).

hooks, b. 2000 *Where We Stand: class matters*, Routledge, London.

House of Lords 2008 The economic impact of immigration, Vol. 1: Report, Select Committee of Economic Affairs, HMSO, London.

Hudson, R. 2000 *Production, Place and Environment*, Prentice-Hall, London.

Hudson, R. 2001 *Producing Places*, Guilford Press, New York.

Hughes, J. 2005 Bringing emotion to work: emotional intelligence, employee resistance and the reinvention of character, *Work, Employment and Society* 19, 603–25.

Hutton, W. 1996 *The State We're In*, Vintage, London.

Ifekwunigwe, J. O. 2006 Recasting 'Black Venus' in the 'new' African diaspora, in K. M. Clarke and D. A. Thomas (eds) *Globalization and Race: transformations in the cultural production of blackness*, Duke University Press, Durham, NC.

Ignatiev, N. 1996 *How the Irish Became White*, Routledge, London.

International Organization for Migration (IOM) 1999 *Migration in Central and Eastern Europe: 1999 review*, IOM, Geneva.

Irwin, S. and Bottero, W. 2000 Market returns? Gender and theories of change in employment relations, *British Journal of Sociology* 51, 261–80.

Isaacs, D. and Poole, M. 1996 Becoming a man and becoming a nurse: three men's stories, *Journal of Gender Studies* 5, 3–47.

Jackson, C. 2002 'Laddishness' as a self-worth protection strategy, *Gender and Education* 9, 117–33.

James, N. 1989 Emotional labour: skill and work in the social regulation of feelings, *Sociological Review* 37, 15–42.

James, N. 1992 Care = organization + physical labour + emotional labour, *Sociology of Health and Illness* 14, 488–509.

Jenkins, S. and Cappellari, L. 2004 Modelling low income transitions, *Journal of Applied Econometrics* 19, 593–610.

Jenkins, S. and Rigg, J. 2001 *The Dynamics of Poverty in the UK*, Department of Work and Pensions Research Report No. 157, DWP, London.

Jensen, T. and Westenholz, A. (eds) 2004 *Identity in the Age of the New Economy*, Edward Elgar, Cheltenham.

Jervis, L. L. 2001 The pollution of incontinence and the dirty work of caregiving in a US nursing home, *Medical Anthropology Quarterly* 15, 84–99.

Johnson, A. 1997 *The Gender Knot: unravelling our patriarchal legacy*, Temple University Press, Philadelphia.

Johnson, A., Wadsworth, J., Wellings, K. and Field, J. 2001 Sexual behaviour in Britain: partnerships, practices and HIV risk behaviours, *The Lancet* 358, 1835–42.

Jones, A. 2008 The rise of global work, *Transactions of the Institute of British Geographers* 33, 12–26.

Kalayaan 1999 *Community Action Against Sex Trafficking: a trainers manual*, Kalayaan, Quezon City, Philippines.

Kang, M. 2003 The managed hand – the commercialization of bodies and emotions in Korean immigrant-owned nail salons, *Gender and Society* 17, 820–39.

Kanter, R. M. 1977 *Men and Women of the Corporation*, Basic Books, New York.

Keep, E. and Mayhew, K. 1999 The assessment: knowledge, skills and competitiveness, *Oxford Review of Economic Policy* 15, 1–15.

Kelly, L. and Regan, L. 2000 *Stopping Traffic: exploring the extent of, and responses to, trafficking in women for sexual exploitation in the UK*, Police Research Series Paper 125, Home Office, London.

Kelly, P. and Moya, E. 2006 Nursing a colonial hangover: race/class/gender and Filipina healthcare workers in Toronto. Paper given at the Annual Conference of the Association of American Geographers, Chicago, 7–11 March (available from the first author at York University, Canada).

Kenway, J. and Fitzclarence, L. 1997 Masculinity, violence and football, *Gender and Education* 9, 117–33.

Kerfoot, D. and Korczynski, M. 2005 Gender and service: new directions for the study of 'front-line' service work, *Gender, Work and Organization* 12, 387–99.

Kim, S.-K. 1997 *Class Struggle or Family Struggle? The lives of women factory workers in South Korea*, Cambridge University Press, Cambridge.

Kimmel, M. 1994 Masculinity as homophobia: fear, shame and silence in the construction of gender identity, in H. Brod and M. Kaufman (eds) *Theorizing Masculinities*, Sage, London.

Knight, D. and McCabe, D. 1998 What happens when the phone goes wild? Staff, stress and spaces for escape in a BPR telephone banking regime, *Journal of Management Studies* 35, 163–94.

Kondo, D. 1990 *Crafting Selves: power, gender and discourse of identity in a Japanese workplace*, University of Chicago Press, Chicago.

Korczynski, M. 2001 The contradictions of service work: the call centre as customer orientated bureaucracy, in. A. Sturdy, I. Grugulis and H. Wilmott (eds) *Customer Service: Empowerment and entrapment*, Palgrave Macmillan, Basingstoke.

Kreiner, G. E., Ashforth, B. E. and Sluss, D. M. 2006 Identity dynamics in occupational dirty work: integrating social identity and system justification perspectives, *Organization Science* 17, 619–36.

Krumal, S. and Geddes, D. 2000 Exploring the dimensions of emotional labour: the heart of Hochschild's work, *Management Communication Quarterly* 14, 8–49.

Lafferty, Y. and McKay, J. 2004 'Suffragettes in satin shorts'? Gender and competitive boxing, *Qualitative Sociology* 27, 249–76.

Lakha, S. 1990 Growth of computer software industry in India, *Economic and Political Weekly* 25, 1, 6 January, p. 53.

Lakha, S. 1994 The new international division of labour and the Indian computer software industry, *Modern Asian Studies* 28, 381–408.

Lash, L. 1994 Reflexivity and its doubles. In U. Beck, A. Giddens and S. Lash (eds) *Reflexive Modernization: politics, tradition and aesthetics in the modern social order*, Polity Press, Cambridge.

Lash, S. 1998 The consequences of reflexivity, in P. Philip (ed.) *Reflexivity and Culture*, Pluto Press, London.

Lash, S., Heelas, P. and Morris, P. (eds) 1996 *De-traditionalization*, Blackwell, Oxford.

Lash, S. and Urry, J. 1987 *The End of Organized Capitalism*, Routledge, London.

Lash, S. and Urry, J. 1994 *Economies of Signs and Space*, Sage, London.

La Valle, I., Finch, S., Nove, A. and Lewin, C. 1999 Parents' demand for childcare, Research Brief 176, DfEE, London.

Lawler, J. 1991 *Behind the Screens : nursing, somology and the problem of the body*, Churchill Livingstone, Melbourne.

Lawler, J. 1997 (ed.) *The Body in Nursing: a collection of views*, Churchill Livingstone, Melbourne.

Lawton, J. 1998 Contemporary hospice care: the sequestration of the unbounded body and 'dirty dying', *Sociology of Health and Illness* 20, 121–43.

Lawton, J. 2000 *The Dying Process: patients' experience of palliative care*, Routledge, London.

Leadbeater, C. 1999 *Living on Thin Air: the new economy*, Penguin, London.

Lee, C. K. 1997 Factory regimes of Chinese capitalism, in A. Ong and D. Nononi (eds) *Ungrounded Empires: the cultural politics of modern Chinese transnationalism*, Routledge, London.

Leidner, R. 1993 *Fast Food, Fast Talk: interactive service work and the routinization of everyday life*, University of California Press, Berkeley.

Lewis, J. 1992 *Women in Britain since 1945*, Blackwell, Oxford.

Lewis, J. 1993 *Women and Social Policies in Europe: gender, family and the state*, Edward Elgar, Cheltenham.

Lewis, J. 2002 Individualization, assumptions about the existence of an adult worker model and the shift towards contractualism, in A. Carling, S. Duncan and R. Edwards (eds) *Analysing Families: morality and rationality in policy and practice*, Routledge, London.

Leyshon, A. and Thrift, N. 1997 *Money/Space: geographies of monetary transformation*, Routledge, London.

Light, A. 2007 *Mrs Woolf and the Servants*, Fig Tree, London.

Llewelyn Davies, M. 1982 *Life As We Have Known It*, Virago, London (originally published in 1939 by the Women's Cooperative Guild).

Lois, J. 2003 *Heroic Efforts: the emotional culture of search and rescue volunteers*, New York University Press, New York.

Lopez, S. 2006 Emotional labour and organized emotional care: conceptualizing nurse home care work, *Work and Occupations* 33, 133–60.

Lupton, B. 2000 Maintaining masculinity: men who do women's work, *British Journal of Management* 11, 233–48.

Lupton, B. 2009 Explaining men's entry into female-dominated occupations, *Gender, Work and Organization* 13, 103–28.

Lupton, D. 1994 *Medicine as Culture: illness, disease and the body in western societies*, Sage, London.

McCall, L. 2001 *Complex Inequality: gender, class and race in the new economy*, Routledge, London.

McCall, L. 2005 The complexity of intersectionality, *Signs: Journal of Women in Culture and Society* 30, 1771–802.

McClintock, A. 1995 *Imperial Leather: race, gender and sexuality in the colonial contest*, Routledge, London.

McCormack, D. 2007 Molecular affects in human geographies, *Environment and Planning A* 39, 359–77.

McCrone, K. 1984 Play up! Play up! And play the game! Sport at the late Victorian girls' public school, *British Studies* 23, 106–34.

Macdonald, C. and Sirianni, C. 1996 *Working in the Service Society*, Temple University Press, Philadelphia.

McDowell, L. 1991 Life without father and Ford: the new gender order of post-Fordism, *Transactions of the Institute of British Geographers* 16, 400–21.

McDowell, L. 1997 *Capital Culture: gender at work in the City*, Blackwell, Oxford.

McDowell, L. 2001 Father and Ford revisited: gender, class and employment change in the new millennium, *Transactions of the Institute of British Geographers* 24, 448–64.

McDowell, L. 2002a Space, place and home, in M. Eagleton (ed.) *Feminist Theory*, Blackwell, Oxford.

McDowell, L. 2002b Masculine discourses and dissonances: Strutting 'lads', protest masculinity and domestic respectability, *Environment and Planning D: Society and Space* 20, 97–119.

McDowell, L. 2003 *Redundant Masculinities? Employment change and white working-class youth*, Blackwell, Oxford.

McDowell, L. 2004 Work, workfare, work/life balance and an ethic of care, *Progress in Human Geography* 28, 1–19.

McDowell, L. 2005 *Hard Labour: the forgotten voices of Latvian migrant 'volunteer' workers*, UCL Press, London.

McDowell, L. 2006a Reconfigurations of gender and class relations, *Antipode* 38, 825–50.

McDowell, L. 2006b Respect, respectability, deference and place: what is the problem with/for working-class boys? *Geoforum* 38, 276–86.

McDowell, L. 2007 Constructions of whiteness: Latvian women workers in postwar Britain, *Journal of Baltic Studies* 38, 1, 85–107.

McDowell, L. 2008 Thinking through work: complex inequalities, constructions of difference and transnational migrants, *Progress in Human Geography* 28, 145–63.

McDowell, L., Batnizky, A. and Dyer, S. 2007 Division, segmentation and interpellation: the embodied labours of migrant workers in a Greater London hotel, *Economic Geography* 83, 1–25.

McDowell, L, Batnitzky, A. and Dyer, S. 2008 Internationalization and the spaces of temporary labour: the global assembly of a local workforce, *British Journal of Industrial Relations* 4: 750–70.

McDowell, L. and Massey, D. 1984 A woman's place? In D. Massey and J. Allen (eds) *Geography Matters!* Cambridge University Press, Cambridge.

McDowell, L., Perrons, D., Ray, K., Fagan, C. and Ward, K. 2005 Women's paid work and moral economies of care, *Social and Cultural Geography* 6, 219–35.

McGovern, P., Smeaton, D. and Hill, S. 2004 Bad jobs in Britain: non-standard employment and job quality, *Work and Occupations* 31, 225–49.

McGregor, J. 2007 Joining the BBC (British Bottom Cleaners): Zimbabwean migrants and the UK care industry, *Journal of Ethnic and Migration Studies* 33, 801–24.

Machin, S. 1999 Wage inequality in the 1970s, 1980s and 1990s, in P. Gregg and J. Wadsworth (eds) *The State of Working Britain*, Manchester University Press, Manchester.

Machin, S. 2008 Rising wage inequality, in *Centre Piece*, Centre for Economic Performance, London School of Economics, London.

McKeganey, N. and Barnard, M. 1996 *Sex Work on the Streets: prostitutes and their clients*, Open University Press, Buckingham.

McNay, L. 2000 *Gender and Agency: reconfiguring the subject in feminist and social theory*, Polity Press, Cambridge.

McOrmand, T. 2004 Changes in working trends over the past decade, *Labour Market Trends*, January, 25–35.

McRae, D. 1992 *Nothing Personal: the business of sex*, Mainstream, Edinburgh.

McRobbie, A. 2004 Notes on 'what not to wear' and post-feminist symbolic violence, in L. Adkins and B. Skeggs (eds) *Feminism after Bourdieu*, Blackwell, Oxford.

Magubane, Z. 2001 Which bodies matter? Feminism, post-structuralism, race and the curious odyssey of the 'Hottentot Venus', *Gender and Society* 15, 816–34.

Mahajan, S. (ed.) 2005 *Input-Output Analysis: 2005*, Office for National Statistics, London.

Mangum, G., Mayall, D. and Nelson, K. 1985 The temporary help industry: a response to the dual internal labour market, *Industrial and Labour Relations Review* 38, 599–611.

Mann, S. 1999 Emotion at work: to what extent are we expressing, suppressing, or faking it? *European Journal of Work and Organizational Psychology* 8, 247–69.

Martin, R. 2001 Local labour markets, in G. Clark, M. Feldman and M. Gertler (eds) *The Oxford Handbook of Economic Geography*, Oxford University Press, Oxford.

Martin, R. and Rowthorne, B. (eds) 1986 *The Geography of Deindustrialization*, Macmillan, London.

Martin, R., Sunley, P. and Wills, J. 1994 Unions and the politics of deindustrialization, *Antipode* 26, 59–76.

Martin, R., Sunley, P. and Wills, J. 1996 *Union Retreat and the Regions*, Jessica Kingsley, London.

Marx, K. 1976 *Capital Vol. 1*, Penguin, London (originally published in 1868).

Mason, G. and Wilson, R. 2003 *Employers Skills Survey: new analyses and lessons learned*, Department for Education and Skills, Nottingham.

Mason, P. 2007 *Live Working, Die Fighting: how the working class went global*, Harvill Secker, London.

Massey, D. 1984 *Spatial Divisions of Labour*, Macmillan, Basingstoke.

Massey, D. 1995 Masculinity, dualisms and high technology, *Transactions of the Institute of British Geographers* 20, 487–99.

Massey, D. 2007 *World City*, Polity Press, Cambridge.

Massey, D. and Meegan, R. 1982 *The Anatomy of Job Loss: the how, why and where of employment decline*, Methuen, London.

May, J., Wills, J., Datta, K., Evans, J., Herbert, J. and McIlwaine, C. 2007 Keeping London working: global cities, the British state and London's new migrant division of labour, *Transactions of the Institute of British Geographers* 32, 151–67.

May, T., Harocopos, A. and Hough, M. 2000 For love or money: pimps and the management of sex work, Police Research Series Paper 134, Home Office, London.

Mellor, P. and Shilling, C. 1997 *Re-forming the Body: religion, community and modernity*, Sage, London.

Mennesson, C. 2000 'Hard' women and 'soft' women: the social construction of identities among female boxers, *International Review for the Sociology of Sport* 35, 21–33.

Menzies Lyth, I. 1959 Social systems as a defence against anxiety, reprinted in I. Menzies Lyth, 1988, *Containing Anxiety in Institutions*, Free Association Press, London.

Merchant, C. 1980 *The Death of Nature*, Harper Collins, New York.

Merrill, H. 2006 *An Alliance of Women: immigration and the politics of race*, University of Minnesota Press, Minneapolis.

Messner, M. 1989 Masculinities and athletics careers, *Gender and Society* 3, 71–88.

Milkman, R. 1987 *Gender at Work: the dynamics of sex segregation during World War II*, University of Illinois Press, Champaign.

Milkman, R. (ed.) 2000 *Organizing Immigrants: the challenge for unions in contemporary California*, Cornell University Press, Ithaca, NY.

Mills, M. B. 2003 Gender and inequality in the global labor force, *Annual Review of Anthropology* 32, 41–62.

Mitchell, K., Marston, S. and Katz, C. (eds) 2004 *Life's Work: geographies of social reproduction*, Blackwell, Oxford.

Mohanty, C. T. 2002 'Under western eyes revisited': feminist solidarity through anti-capitalist struggles, *Signs: Journal of Women in Culture and Society* 28, 499–535.

Moi, T. 2005 *Sex, Gender and the Body: the student edition of What is a Woman?* Oxford University Press, Oxford.

Monaghan, L. 2002 Regulating 'unruly' bodies: work tasks, conflict and violence in Britain's night-time economy, *British Journal of Sociology* 53, 403–29.

Moreira, T. 2004 Coordination and embodiment in the operating room, *Body and Society* 10, 109–29.

Morgan, K. P. 1994 *Women and the Knife: cosmetic surgery and the colonization of women's bodies*, Westview Press, Boulder, CO.

Morris, J. A. and Feldman, D. C. 1996 The dimensions, antecedents and consequences of emotional labour, *Academy of Management Review* 21, 986–1010.

Mount, F. 2004 *Mind the Gap: the new class divide in Britain*, Short Books, London.

Mulholland, K. 2002 Gender, emotional labour and telemarketing in a call centre, *Personnel Review* 31, 283–303.

Naisbitt, J., Naisbitt, N. and Philips, D. 2001 *High Tech – High Touch: technology and our accelerated search for meaning*, Nicholas Brealey Publishing, London.

Newman, K. 1999 *No Shame in My Game: the working poor in the inner city*, Vintage Books and Russell Sage Foundation, New York.

Newman, K. 2006 *Chutes and Ladders: navigating the low wage labor market*, Russell Sage Foundation, New York and Harvard University Press, Cambridge, MA.

Nolan, P. and Slater, G. 2008 Living on hot air: myths and realities in the 'new economy', paper given at the conference 'Transforming Work', St John's College Research Centre, Oxford, September (paper available from the first author at the University of Leeds).

Novarra, V. 1980 *Men's Work, Women's Work*, Marion Boyars, London.

Oakley, A. 1974a *The Sociology of Housework*, Martin Robertson, London.

Oakley, A. 1974b *Housewife*, Allen Lane, London.

Oates, J. C. 1987 *On Boxing*, Doubleday, Garden City, NY.

O'Brien, K., Hobbs, D. and Westmarland, L. 2007 Negotiating violence and gender: security and the night time economy in the UK, in S. Body-Gendrot and P. Spierenburg (eds) *Cultures of Violence in Europe: historical and contemporary perspectives*, Springer, New York.

O'Connell Davidson, J. 1995 The anatomy of 'free choice' prostitution, *Gender, Work and Organization* 2, 1–10.

O'Connell Davidson, J. 1996 Prostitution and the contours of control, in J. Weeks and J. Holland (eds) *Sexual Cultures: commodities, values and intimacy*, Macmillan, London.

O'Connell Davidson, J. 1998 *Prostitution, Power and Freedom*, Polity Press, Cambridge.

Office of National Statistics (ONS) 2005 *National Statistics: UK Snapshot*, HMSO, London.

Office of National Statistics (ONS) 2007 *Social Trends 37: 2007*, HMSO, London.

Office of National Statistics (ONS) 2008a *Labour Market Snapshot*, HMSO, London.

Office of National Statistics (ONS) 2008b *Social Trends 38: 2008*, HMSO, London.

Oishi, N. 2005 *Women in Motion: globalization, state policies and labour migration in Asia*, Stanford University Press, Palo Alto, CA.

O'Neill, M. 1996 The aestheticization of the whore in contemporary society: desire, the body, self and society, paper presented to the Body and Organization Workshop, Keele, September.

O'Neill, M. 2001 *Prostitution and Feminism*, Polity Press, Cambridge.

Ong, A. 1987 *Spirits of Resistance and Capitalist Discipline: factory women in Malaysia*, State University of New York Press, Albany.

Ong, A. 1991 The gender and labor politics of postmodernity, *Annual Review of Anthropology* 20, 279–309.

Ong, A. 1996 Cultural citizenship as subject-making: immigrants negotiate racial and cultural boundaries in the US, *Current Anthropology* 37, 737–62.

Ong, A. 2003 *Buddha is Hiding*, University of California Press, Berkeley.

Organization for Economic Cooperation and Development (OECD) 2008 *Growing Unequal: income distribution and poverty in OECD countries*, Directorate for Employment, Labour and Social Affairs, Paris.

Ouellet, L. 1994 *Pedal to the Metal: the working lives of truckers*, Temple University Press, Philadelphia.

Pahl, R. 1988 *On Work*, Blackwell, Oxford.

Palmer, P. 1989 *Domesticity and Dirt: housewives and domestic servants in the United States, 1920–1945*, Temple University Press, Philadelphia.

Panitch, L. and Leys, C. (eds) 2000 *Working Classes, Global Realities*, Merlin Press, London.

Park, R. and Burgess, E. W. 1921 *Introduction to the Science of Sociology*, Chicago University Press, Chicago.

Park, R. and Burgess, E. W. 1984 *The City: suggestions for investigation of human behaviour in the urban environment*, University of Chicago Press, Chicago (first published 1925).

Park, R., Burgess, E. W. and McKenzie, R. 1967 *The City*, University of Chicago Press, Chicago.

Parreñas, R. 2001 *Servants of Globalization*, Stanford University Press, Palo Alto, CA.

Pateman, C. 1988 *The Sexual Contract*, Polity Press, Cambridge.

Paul, K. 1997 *Whitewashing Britian: race and citizenship in the postwar era*, Cornell University Press, Ithaca, NY.

Payne, J. 2006 Emotional labour and skill: a re-appraisal, Skope Issues Paper 10, University of Warwick.

Pearson, G. 1983 *Hooligan: a history of respectable fears*, Macmillan, London.

Peck, J. 1996 *Work/Place*, Guilford Press, New York.

Peck, J. 2000 *Workfare States*, Guilford Press, New York.

Peck, J. and Theodore, N. 2001 Contingent Chicago: restructuring the spaces of temporary labour, *International Journal of Urban and Regional Research* 25, 471–96.

Peck, J. and Theodore, N. 2002 Temped out? Industry rhetoric, labour regulation and economic restructuring in the temporary staffing industry, *Economic and Industrial Democracy* 23, 143–75.

Peck, J., Theodore, N. and Ward, K. 2005 Constructing markets for temporary employment: employment liberalization and the internationalization of the staffing industry, *Global Networks* 5, 3–26.

Perkins, R. 1991 *Working Girls: prostitutes, their life and social control*, Australian Institute of Criminology, Canberra.

Perrons, D., Fagan, C., McDowell, L., Ward, K. and Ray, K. (eds) 2006 *Gender Divisions and Working Time in the New Economy*, Edward Elgar, Cheltenham.

Philpin, S. M. 1999 The impact of 'Project 2000' educational reforms on the occupational socialization of nurses: an exploratory study, *Journal of Advanced Nursing* 29, 1326–31.

Pierce, J. 1995 *Gender Trials: emotional lives in contemporary law firms*, University of California Press, Berkeley.

Piore, M. 1979 *Birds of Passage: migrant labour and industrial societies*, Cambridge University Press, Cambridge.

Piore, S. and Sabel, C. 1984 *The Second Industrial Divide*, Basic Books, New York.

Pitt, K. 2002 Being a new capitalist mother, *Discourse and Society* 13, 251–67.

Pohl, F. 1977 *Gateway*, Gollanz, London.

Polanyi, M. 1967 *The Tacit Dimension*, Routledge and Kegan Paul, London.

Pratt, G. 1997 Stereotypes and ambivalence: the construction of domestic workers in Vancouver, BC, *Gender, Place and Culture* 4, 159–77.

Pratt, G. 1999 From registered nurse to registered nanny: discursive geographies of Filipina domestic workers in Vancouver, BC, *Economic Geography* 75, 215–36.

Pratt, G. 2003 Valuing childcare: troubles in suburbia, *Antipode* 35, 581–602.

Pratt, G. 2004 *Working Feminism*, Edinburgh University Press, Edinburgh.

Prince's Trust, 2007 The cost of exclusion. Report commissioned from the Centre for Economic Performance, LSE, Prince's Trust, London.

Pringle, R. 1988 *Secretaries Talk: sexuality, power and work*, Verso, London.

Pringle, R. 1998 *Sex and Medicine: gender, power and authority in the medical profession*, Cambridge University Press, Cambridge.

Probyn, E. 2004 Teaching bodies: affects in the classroom, *Body and Society* 10, 21–43.

Prus, R. and Vassilakopoulos, S. 1979 Desk clerks and hookers: hustling in a 'shady' hotel, *Journal of Contemporary Ethnography* 8, 52–71.

Pryke, M. 1991 An international city going global: spatial change in the City of London, *Environment and Planning D: Society and Space* 9, 197–222.

Putnam, R. 2000 *Bowling Alone: the collapse and revival of American community*, Simon and Schuster, New York.

Pyke, K. 1996 Class-based masculinities: the interdependence of gender, class and interpersonal power, *Gender and Society* 10, 527–49.

Radcliffe, S. 1990 Ethnicity, patriarchy and incorporation into the nation: female migrants as domestic servants in Peru, *Environment and Planning D: Society and Space* 8: 379–93.

Reay, D. 2002 Shaun's story: troubling discourses of white working-class masculinities, *Gender and Education* 14, 221–34.

Reich, R. 1991 *The Work of Nations: preparing ourselves for 21st century capitalism*, Knopf, New York.

Richardson, R., Belt, V. and Marshall, N. 2000 Taking calls in Newcastle: the regional implications of the growth in call centres, *Regional Studies* 34, 357–69.

Rifkin, J. 1995 *The End of Work*, Tarcher, New York.

Ritzer, G. 1999 *Enchanting a Disenchanted World: revolutionizing the means of consumption*, Pine Forge Press, London.

Robinson, J. 2006 *Ordinary Cities: between modernity and development*, Routledge, London.

Rodrigues, M. (ed.) 2005 The new knowledge economy in Europe: a strategy for international competitiveness and social cohesion, Edward Elgar, Cheltenham.

Roediger, D. 1999 *The Wages of Whiteness: race and the making of the American working class*, Verso, London.

Rogers, J. 1995 'Just a temp': experience and structure of alienation in temporary clerical employment, *Work and Occupations* 22, 137–66.

Romero, M. 1992 *Maid in the USA*, Routledge, London.

Rosaldo, R. 1993 *Culture and Truth: the remaking of social analysis*, Beacon Press, Boston.

Rosewarne, S. 2001 Globalization, migration and labour market formation – labour's challenge, *Capital, Nature and Society* 12, 71–84.

Said, E. 1994 *Culture and Imperialism*, Knopf, New York.

Salih, S. 2002 *Judith Butler*, Routledge, London.

Salzinger, L. 2003 *Genders in Production: making workers in Mexico's global factories*, University of California Press, Berkeley.

Samuel, R. (ed.) 1977 *Miners, Quarrymen and Saltworkers*, Routledge and Kegan Paul, London.

Sanders, T. 2005 'It's just acting': sex workers' strategies for capitalizing on sexuality, *Gender, Work and Organization*, 12, 319–42.

Sass, J. S. 2000 Emotional labour as cultural performance: The communication of caregiving in a non-profit nursing home, *Western Journal of Communication* 64, 330–58.

Sassen, S. 1988 *The Mobility of Labour and Capital*, Cambridge University Press, Cambridge.

Sassen, S. 2001 *The Global City: New York, London and Tokyo*, Princeton University Press, Princeton, NJ.

Savage, L. 1998 Geographies of organizing: justice for janitors in Los Angeles, in A. Herod (ed.) *Organizing the Landscape: geographical perspectives on labor unions*, University of Minnesota Press, Minneapolis.

Savage, M. 2000 *Class Analysis and Social Transformation*, Open University Press, Buckingham.

Sayer, A. 2005 *The Moral Significance of Class*, Cambridge University Press, Cambridge.

Scannell, V. 1960 *The Big Chance*, Chatto and Windus, London.

Scott, J. 1985 *Weapons of the Weak: everyday strategies of peasant resistance*, Yale University Press, New Haven, CT.

Seifert, A. M. and Messing, K. 2006 Cleaning up after globalization: an ergonomic analysis of work activity of hotel cleaners, *Antipode* 38, 557–78.

Sennett, R. 1998 *The Corrosion of Character: the personal consequences of work in the new capitalism*, W. W. Norton, New York.

Sennett, R. 2006 *The Culture of the New Capitalism*, Yale University Press, New Haven, CT.

Seymour, D. and Sandiford, P. 2005 Learning emotion rules in service organizations: socialization and training in the UK public-house sector, *Work, Employment and Society* 19, 547–64.

Sharma, U. and Black, P. 2001 Look good, feel better: beauty therapy as emotional labour, *Sociology* 35, 913–31.

Sharpley-Whiting, T. D. 1999 *Black Venus: sexualized savages, primal fears and primitive narratives in French*, Duke University Press, Durham, NC.

Shilling, C. 1993 *The Body and Social Theory*, Sage, London.

Silvestro, R., Fitzgerald, L., Johnston, R. and Voss, C. 1992 Towards a classification of service process, *International Journal of Services Industry Management* 3, 62–75.

Simpson, M. 1994 *Male Impersonators: men performing masculinity*, Cassell, London.

Simpson, R. 1997 Have times changed? Career barriers and the token woman manager, *British Journal of Management* 8, 121–9.

Simpson, R. 2004 Masculinity at work: the experiences of men in female dominated occupations, *Work, Employment and Society* 18, 349–68.

Skeggs, B. 1997 *Formations of Class and Gender: becoming respectable*, Sage, London.

Skeggs, B. 2004a *Class, Self, Culture*, Routledge, London.

Skeggs, B. 2004b Context and background: Pierre Bourdieu's analysis of class, gender and sexuality, in L. Adkins and B. Skeggs (eds) *Feminism after Bourdieu*, Blackwell, Oxford.

Smith, A. 1986 *The Wealth of Nations, Books I to III*, Penguin, London (originally published in 1776).

Smith, D. 1987 *The Everyday World as Problematic: a feminist sociology*, Northeastern University Press, Boston.

Smith, D. 1990 *The Conceptual Practices of Power: a feminist sociology of knowledge*, Northeastern University Press, Boston.

Smith, D. 1991 *Texts, Facts and Femininity: exploring the relations of ruling*, Routledge, Boston.

Smith, M. 2005 The incidence of new forms of employment in service activities, in G. Bosch and S. Lehndorff (eds) *Working in the Service Sector: a tale of different worlds*, Routledge, London.

Smith, P. 1992 *The Emotional Labour of Nursing: how nurses care*, Macmillan, London.

Smith, P. and MacKintosh, M. 2007 Profession, market and class: nurse migration and the remaking of division and disadvantage, *Journal of Clinical Nursing* 16, 2213–20.

Social Exclusion Unit 1998 *Bringing Britain Together: a national strategy for neighbourhood renewal*, Cmnd 4045, HMSO, London.

Soja, E. 1989 *Postmodern Geographies: the reassertion of space in critical social theory*, Verso, London.

Spanger, M. 2002 Black prostitutes in Denmark, in S. Throbeck and B. Paatanaik (eds) *Transnational Prostitution: changing global patterns*, Zed Books, London.

Spence, L. 2005 *Country of Birth and Labour Market Outcomes in London: an analysis of labour force and census data*, Greater London Council, London.

Spivak, G. C. 1999 *A Critique of Postcolonial Reason*, Harvard University Press, Cambridge, MA.

Spring Rice, M. 1939 *Working-Class Wives: their health and conditions*, Penguin, London.

Stacey, C. L. 2005 Finding dignity in dirty work: the constraints and rewards of low-wage home care labour, *Sociology of Health and Illness* 27, 831–54.

Staples, R. 2006 *Exploring Black Sexuality*, Rowman and Littlefield, Lanham, MD.

Steedman, C. 2007 *Master and Servant: love and labour in the English industrial age*, Cambridge University Press, Cambridge.

Steill, B. and England, K. 1999 Jamaican domestics, Filipina housekeepers and English nannies: representations of Toronto's foreign domestic workers, in J. Momsen (ed.) *Gender, Migration and Domestic Service*, Routledge, London.

Steinberg, R. and Figart, D. 1999 Emotional labor since *The Managed Heart, Annals of the American Academy of Political and Social Science* 561, 8–26.

Stewart, M. 2007 The inter-related dynamics of unemployment and low-wage employment, *Journal of Applied Econometrics* 22, 511–31.

Stone, K. 2004 *From Widgets to Digits: employment regulation for the changing workplace*, Cambridge University Press, Cambridge.

Storper, M. and Walker, R. 1989 *The Capitalist Imperative: territory, technology and industrial growth*, Blackwell, Oxford.

Storrie, D. 2002 *Temporary Agency Work in the European Union*, European Foundation for the Improvement of Living and Working Conditions, Office for Official Publications of the European Communities, Luxembourg.

Strathern, M. 1987 The limits of auto-anthropology, in A. Jackson (ed.) *Anthropology at Home*, Tavistock, London.

Strossen, N. 1996 *Defending Pornography*, Abacus, London.

Sugden, J. 1987 The exploitation of disadvantage: the occupational subculture of the boxer, in J. Horne, D. Jary and A. Tomlinson (eds) *Sport, Leisure and Social Relations*, Routledge and Kegan Paul, London.

Sunley, P., Martin, R. and Nativel, C. 2001 Mapping the New Deal: local disparities in the performance of Welfare-to-Work, *Transactions of the Institute of British Geographers* 26, 484–512.

Tanenbaum, L. 1994 The politics of porn: forced arguments, *In These Times*, 7 March, 17–20.

Taylor, J. S. 2001 Dollars are a girl's best friend? Female tourists' sexual behaviour in the Caribbean, *Sociology* 35, 749–64.

Taylor, P. and Bain, F. 1999 'An assembly line in the head': work and employee relations in the call centre, *Industrial Relations Journal* 30, 101–17.

Teather, D. 2008 Special report: executive pay, *Guardian*, 11 September, pp. 28–9 and 12 September, p. 30.

Theodore, N. and Peck, J. 2002 The temporary staffing industry: growth imperatives and limits to contingency, *Economic Geography* 78, 462–93.

Thrift, N. 2005 *Knowing Capitalism*, Sage, London.

Thrift, N. 2007 *Non-Representational Theory: space, politics, affect*, Routledge, London.

Thrift, N. Leyshon, A. and Daniels, P. 1987 'Sexy greedy': the new international financial system, the City of London and the South East of England, Working Papers in Producer Services 8, University of Bristol.

Tilly, C. 1998 *Durable Inequality*, University of California Press, Berkeley.

Tilly, C. and Tilly, C. 1998 *Work Under Capitalism*, Westview Press, Boulder, CO.

Titmuss, R. 1997 *The Gift Relationship: from human blood to social policy*, New Press, New York (revised and expanded edition; originally published 1970).

Tolia-Kelly, D. 2006 Affect – an ethnocentric encounter? Exploring the 'universalist' imperative of emotional/affective geographies, *Area* 38, 213–17.

Toynbee, P. 2003 *Hard Work: life in low-pay Britain*, Bloomsbury Press, London.

Travis, A. 2007 11,000 illegal migrants licensed to work as private security guards, *Guardian*, 14 December, p. 2.

Treanor, J. 2003 Furse hits back at 'frivolous' media, *Guardian*, 27 February, p. 26.

Tseelon, E. 1995 *The Masque of Femininity*, Sage, London.

Tufts, S. 2006 'We make it work': the cultural transformation of hotel workers in the city, *Antipode* 38, 350–73.

Turner, B. 1996 *The Body and Society: explorations in social theory*, Sage, London.

Turner, B. 2008 *The Body and Society: explorations in social theory*, 3rd edn, Sage, London.

Turner, J. H. and Stets, J. E. 2006 Sociological theories of human emotions, *Annual Review of Sociology* 32, 25–52.

Twigg, J. 2000a Carework as a form of bodywork, *Ageing and Society* 20, 389–411.

Twigg, J. 2000b *Bathing: the body in community care*, Routledge, London.

Twigg, J. 2006 *The Body in Health and Social Care*, Palgrave Macmillan, Basingstoke.

Tyler, M. and Abbott, P. 1998 Chocs away: weight watching in the contemporary airline industry, *Sociology* 14, 433–50.

Urry, J. 1990 *The Tourist Gaze: travel and leisure in contemporary societies*, Sage, London.

Urry, J. 2000 *Sociology Beyond Societies: mobilities for the twenty-first century*, Routledge, London.

Urry, J. 2004 Connections, *Environment and Planning D: Society and Space* 22, 27–37.

Van Maanen, J. 1991 The smile factory: work at Disneyland, in P. Frost, L. Moore, M. Louis, C. Lundberg and J. Martin (eds) *Reframing Organizational Culture*, Sage, London.

Vertovec, S. 2006 The emergence of super-diversity in Britain, Working Paper 25, Centre for Migration, Policy and Society (COMPAS), University of Oxford (available from the author).

Vosko, L. 2001 *Precarious Work*, Toronto University Press, Toronto.

Vosko, L. (ed.) 2006 *Precarious Employment: understanding labour market insecurity in Canada*, McGill-Queens University Press, London.

Walby, S. 1986 *Patriarchy at Work*, Polity Press, Cambridge.

Walby, S. 1990 *Theorizing Patriarchy*, Blackwell, Oxford.

Waldinger, R. 1992 Taking care of the guests: the impact of immigrants on services – an industry case study, *International Journal of Urban and Regional Research* 16, 97–113.

Waldinger, R. and Lichter, M. I. 2003 *How the Other Half Works: immigrants and the social organization of labour*, University of California Press, Berkeley.

Walkowitz, J. 1980 *Prostitution and Victorian Society*, Cambridge University Press, Cambridge.

Walkowitz, J. 1992 *City of Dreadful Delight: narratives of sexual danger in late Victorian London*, Virago, London.

Waquant, L. 1992 The social logic of boxing in Black Chicago: toward a sociology of pugilism, *Sociology of Sport Journal* 9, 221–54.

Waquant, L. 1995a Pugs at work: bodily capital and bodily labour among professional boxers, *Body and Society* 1, 65–94.

Waquant, L. 1995b The pugilistic point of view: how boxers think and feel about their trade, *Theory and Society* 24, 489–535.

Waquant, L. 2003 *Body and Soul: notebooks of an apprentice boxer*, Oxford University Press, Oxford.

Ward, H., Mercer, C. and Wellings, K. 2005 Who pays for sex? An analysis of the increasing prevalence of female commercial sex contacts among men in Britain, *Sexually Transmitted Infections* 81, 467–71.

Ward, K. 2003 UK temporary staffing: industry structure and evolutionary dynamics, *Environment and Planning A* 35, 889–909.

Ward, K. 2004 Going global? Internationalization and diversification in the temporary staffing industry, *Journal of Economic Geography* 4, 251–73.

Ward, K. 2007 Thinking geographically about work, employment and society, *Work, Employment and Society* 21, 265–76.

Waring, M. 1988 *If Women Counted: a new feminist economics*, Harper Collins, London.

Webb, L. and Elms, J. 1994 Social workers and sex workers, in R. Perkins, G. Prestage, R. Sharp and F. Lovejoy (eds) *Sex Work and Sex Workers*, University of New South Wales Press, Sydney.

Weber, M. 2002 *The Protestant Ethic and the Spirit of Capitalism*, Penguin, New York (originally published 1904–5).

Webster, W. 1998 *Imagining Home: gender, 'race' and national identity 1945–64*, UCL Press, London.

Weinberg, S. and Arond, H. 1969 The occupational culture of the boxer. In J. W. Loy and G. S. Kenyon (eds) *Sport, Culture and Society*, Macmillan, New York.

West, C. and Zimmerman, D. 1987 Doing gender, *Gender and Society* 1, 125–51.

Westwood, S. 1984 *All Day, Every Day*, Pluto Press, London.

Westwood, S. and Phizaklea, A. 2000 *Transnationalism and the Politics of Belonging*, Routledge, London.

Whitehead, S. and Barrett, F. (eds) 2001 *The Masculinities Reader*, Polity Press, Cambridge.

Williams, C. 1993 *Doing Women's Work: men in non-traditional occupations*, Sage, London.

Williams, C. 2003 Sky service: the demands of emotional labour in the airline industry, *Gender, Work and Organization* 10, 513–50.

Williams, C. 2006 *Inside Toyland: working, shopping and social inequality*, University of California Press, Berkeley.

Willis, P. 1977 *Learning to Labour: how working-class kids get working-class jobs*, Saxon House Press, Westmead, Sussex.

Willis, P. 2000 *The Ethnographic Imagination*, Polity Press, Cambridge.

Wills, J. 2004 Organizing the low paid: East London's living wage campaign as a vehicle for change, in G. Healy, E. Heery, P. Taylor and W. Brown (eds) *The Future of Worker Representation*, Palgrave Macmillan, Basingstoke.

Wills, J. and Waterman, P. 2001 *Space, Place and the New Labour Internationalisms*, Blackwell, Oxford.

Wilson, E. 1991 *The Sphinx in the City: urban life, the control of disorder and women*, Virago, London.

Winder, R. 2004 *Bloody Foreigners*, Little Brown, London.

Winlow, S. 2001 *Bad Fellas: crime, tradition and new masculinities*, Berg, Oxford.

Wintour, P. 2008a Brown offers pay commission for 1m agency workers, *Guardian*, 14 February, p. 11.

Wintour, P. 2008b Agency and temporary workers win rights to deal, *Guardian*, 21 May, p. 15.

Wolfe, T. 1987 *Bonfire of the Vanities*, Cape, London.

Wolkowitz, C. 2002 The social relations of body work, *Work, Employment and Society* 16, 497–510.

Wolkowitz, C. 2006 *Bodies at Work*, Sage, London.

Woodward, R. 2004 *Military Geographies*, Blackwell, Oxford.

Woodward, R. and Winter, T. 2007 *Sexing the Soldier: the politics of gender and the contemporary British army*, Routledge, London.

Wright, M. 1997 Crossing the factory frontier: place and power in the Mexican maquiladora, *Antipode* 29, 278–302.

Wright, M. 1999 The politics of relocation: gender, nationality and value in the macquiladoras, *Environment and Planning A* 31, 1601–17.

Wright, M. 2006 *Disposable Women and Other Myths of Global Capitalism*, Routledge, London.

Wright Mills, C. 1953 *White Collar Work: the American middle classes*, Oxford University Press, Oxford.

Xiang, B. 2006 *Global Body Shopping: an Indian labour system in the information technology industry*, Princeton University Press, Princeton, NJ.

Young, I. M. 1990 *Justice and the Politics of Difference*, Princeton University Press, Princeton, NJ.

Young, I. M. 2005 *On Female Body Experience: 'throwing like a girl' and other essays*, Oxford University Press, Oxford.

Young, R. 1995 *Colonial Desire: hybridity in theory, culture and race*, Routledge, London.

Zapf, D. and Holz, M. 2006 On the positive and negative effects of emotion work in organizations, *European Journal of Work and Organizational Psychology* 15, 1–28.

Zatz, N. 1997 Sex work/sex act: law, labour and desire in constructions of prostitution, *Signs* 22, 277–308.

Index

Page numbers in italics denote illustrations, figures or tables

Mooney, A. 88, 89, 93
Morris, J. 49
mortgage providers 34
Moss, P. 88, 89, 93
motherhood 87, 88–9, 93
Mount, F. 71
multinational corporations 211
MySpace 224

Naisbitt, J. 44
Naisbitt, N. 44
nannies 15, 89, 90, 91, 96, 97
Narula, R. 92
National Health Service 210
national stereotyping 202, 206
Nationality, Immigration
    and Asylum Act 123
NEET category 43
Nelson, J. A. 175
neoliberal employment policies 4
network society 5, 32, 42–3
New York City
    fast food workers 144, 197–8
    humiliation/respect 195
    young men 192
Newman, K. 144, 192, 197–8, 199
Nigerian migrant workers 118
Nightingale, F. 163
Nolan, P. 37, *38*
Northern Rock 217
nursing
    disgust 166
    distancing behaviour 166
    and doctors 161
    from home to hospital 163
    men in 56
    non-UK born 217
    as semi-profession 162
    stress 165–6, 183
nursing assistant study 179–84
nursing auxiliaries 166–7

Oates, J. C. 137, 144
O'Brien, K. 149, 150
Observer 119–20
occupations, gendered 52–3
O'Connell Davidson, J. 110, 112

Odyssey (Homer) 129, 130
OECD data 30, 41
Office of National Statistics (ONS) 27,
    30, *39*, *170*
Olympic swimmers study 144
O'Neill, M. 107, 109
Ong, A. 73, 74, 184, 199
organizations 52, 61
Others
    desire 75
    exclusion 156
    exoticism 116
    labour division 52
    live-in employees 91–3
    racialized 71–6
    stereotyping 63, 184
    white, middle-class men 133
Ouellet, L. 135

Pahl, R. 3, 11, 228
pain 145–6
Park, R. 134
participant observation 134, 148–51,
    153–6, 179–84
part-time employment 3, 41, 88
Pateman, C. 105–6, 107
patriarchal assumptions 106
Peck, J. 211
pension funds 213, 228
people of colour 71–2, 116–19, 180
    see also skin colour; *specific groups*
people skills 35
performance
    body 13, 146
    deferential 196, 199
    embodiment 49, 50, 156–7
    femininity 55–6
    gender 51, 131
    masculinity 55–6, 131, 192
    transgressive 56
    virtual 223
performativity 33, 55–6, 72, 131, 203
Perkins, R. 106
personal services 37, 47
personality work 50–1
Philips, D. 44
Phizaklea, A. 118, 227